POPULAR SCIENCE®
THE FUTURE NOW

THE ULTIMATE DIY
TECH
UPGRADES
GUIDE

EDITED BY DOUG CANTOR

POPULAR SCIENCE® THE FUTURE NOW

THE ULTIMATE DIY

TECH UPGRADES GUIDE

weldon**owen**

contents

FOREWORD

Anyone can make anything. That is the lesson of the 140 years that *Popular Science* has been in print. A determined person, working on the weekends, can escape gravity, break the speed of sound, or create a new means of communicating across great distances. And that inventive process begins by tearing things apart and rebuilding them again.

I cannot claim to have been the kind of kid who did that. My instincts around technology were always to keep my belongings clean, dry, and otherwise in perfect working order, not to dismantle or modify them. But my years at *Popular Science* have taught me that that instinct is the wrong one. There is simply too much technology at our disposal not to mess around with it. And when a determined person brings his or her inventive instincts to bear on the gadgets and gizmos that fill our lives, great things can result.

But this book isn't necessarily about building great things. It's about messing around, usually for the simple fun of it. The projects on these pages are based, in part, on the years we've spent pursuing the lone, sometimes crazed hackers who don't just modify technology but blow it apart, just to be able to say they've done it.

I spent an afternoon with our staff photographer, John Carnett, who in his off time had replaced the motor on a four-wheel ATV with a jet engine. The thing required an elaborate start-up procedure and ear protection just to get it rolling, and as he drove me through his Philadelphia neighborhood in it, I cringed to imagine the incredible racket we were making, essentially that of a 747 taxiing past. And yet throughout the process, oblivious to the enemies he was making among the local parents, dogs, and nappers, John wore a look of joy and pride that had

nothing to do with serving humanity or inventing something new. He'd just tricked something out—hacked it—in his own way, and in doing so had made his mark on the universe, not to mention on the neighborhood noise ordinances.

It's in that spirit that we, and especially our tireless senior editor Doug Cantor, bring you this entertaining collection of projects. We did it for the hell of it.

Jacob Ward

Jacob Ward
Editor-in-Chief
Popular Science

INTRODUCTION

I was an unlikely candidate to become the editor of How 2.0, *Popular Science*'s do-it-yourself column. I've always been reasonably handy, but when it came to real hands-dirty, open-things-up-and-rearrange-the-parts hacking skills, I was a complete novice.

So I got into the DIY world the same way an experienced DIYer would work on a project: I did some research, talked to seasoned tinkerers, and then dove in. Early on I managed to build a tiny flashlight, hack my cell phone's firmware, and make a pair of bookends from old CDs, all without causing too much damage. Over time I found that with a few hours, a small pile of parts from Radio Shack, and a little patience, I could build some really cool stuff.

Editing How 2.0 has also given me a window into the vast community of smart, dedicated people who have found about a million uses for things like solenoid valves and Arduino microcontrollers. The breadth of their innovations is truly astonishing, and that's what *Popular Science* has tried to show in the pages of How 2.0 each month. We've featured projects ranging from a remote-controlled helicopter only 2 inches (5 cm) high to a remote-controlled bomber with a 20-foot (60-m) wingspan; from a portable solar-powered gadget charger to a 200-pound (90-kg) solar-powered 3-D printer; from a robot built from a toothbrush head to one that can mix and serve cocktails.

The projects in this book, selected from *The Big Book of Hacks*, are all about customizing the electronics in your life—

sometimes to make them hyperfunctional, and always to make them way more awesome. A few of the featured hacks are audacious things that almost no one could (or probably should) try. Most, though, are easier to replicate. Some take only a few minutes and require little more than gluing parts together.

So if you've never attempted to make anything before in your life, this book provides plenty of ways to start. From there you can take on some of the more challenging projects and develop new skills. Eventually, you might actually find yourself advancing your projects far beyond the versions in the book.

Whatever your skill level and area of interest, I encourage you to roll up your sleeves and (safely) give one of these projects a shot. At times you may get frustrated or even break something. But ultimately you'll be surprised by what you can make, hack, tweak, improve, and transform—and by how much fun you'll have doing it.

Douglas Cantor

Doug Cantor
Senior Editor
Popular Science

HOW TO USE THIS BOOK

So you want to hack stuff—to tear it apart, put it back together with other components, and make it new. We at *Popular Science* salute you, and we've put together these projects to get you started. Many of them come from our popular How 2.0 column, and many come from amazingly inventive individuals out there making cool stuff. (Check out the "Thanks to Our Makers" section for more info.) Before starting a project, you can look to the following symbols to decode what you're facing.

If you're just breaking in your screwdriver and have never even heard the word *microcontroller*, try out these projects first. Designed to be doable within five minutes—give or take a few seconds, depending on your dexterity—and to make use of basic household items, these tech crafts are the perfect starting ground for the newbie tinkerer.

Popular Science has been doing DIY for a very long time—almost as long as the 140 years the magazine has been in print. Occasionally this book shares a DIY project from our archives so you can try your hand at the hilarious retro projects your grandfather and grandmother built back before, say, television or smartphones.

Everyone loves a good success story—tales of everyday individuals who created something so wild that it makes you say . . . well, "You built WHAT?!" You'll find several of these stories throughout the book, and it is our hope that they will inspire you to take your projects to the next level.

BUILD IT!

These are the big ones—the ambitious projects that you'll want to sink some real time and cash into, and that will challenge your skills as a builder. How much time and cash, you ask? And just how challenging? The helpful rubric below will give you an idea.

COST

$ = UNDER $50

$$ = $50–$300

$$ = $300–$1,000

$$$$ = $1,000 AND UP

TIME

☺ UNDER 1 HOUR

☺ ☺ 2–5 HOURS

☺ ☺ ☺ 5– 10 HOURS

☺ ☺ ☺ ☺ 10 HOURS AND UP

DIFFICULTY

● ○ ○ ○ ○ One step up from 5 Minute Projects, these tutorials require basic builder smarts, but no electronics or coding wizardry.

● ● ○ ○ ○ Slightly advanced building skills are a must for these projects. Turn on your common sense, and troubleshoot as you go—it's part of the fun.

● ● ● ○ ○ If you see three dots, it means an activity demands low-level electronics and coding skills, or it's pretty rigorous, construction-wise.

● ● ● ● ○ You likely need serious circuitry know-how and code comprehension to do these projects. That, or be prepared to sweat with heavy-duty assembly.

● ● ● ● ● Hey, if you're reading this book, we figure you like challenges. And projects marked with five dots are sure to deliver just that.

WARNING
If you see this symbol, we mean business. Several of the projects in this book involve dangerous tools, electrical current, flammables, potentially harmful chemicals, and recreational devices that could cause injury if misused. So remember: With great DIY comes great responsibility. Use your head, know your tools (and your limitations), always wear safety gear, and never employ your hacking prowess to hurt others. (See our Disclaimer for more information about how *Popular Science* and the publisher are not liable for any mishaps.)

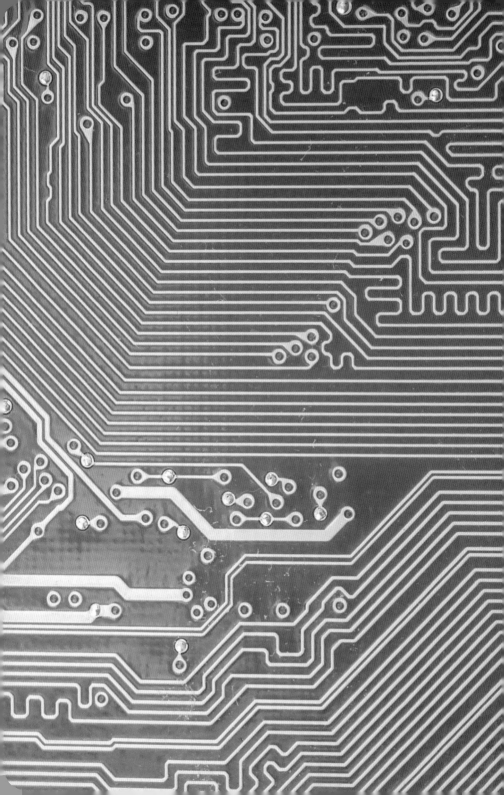

BASIC BUILDING SMARTS

001 PUT TOGETHER A SOLDERING KIT

Soldering is playing with fire, or at least with hot metal. So you need the right tools.

If you're working on electronics projects, you'll need to connect metal objects with a strong, conductive bond. Soldering is the way to go. You heat pieces of metal with a soldering iron, then join them together using a molten filler, or solder.

SOLDER This is the good stuff—the material you'll melt to connect metals. Traditionally, solder was a mix of tin and lead, but these days look for lead-free types to avoid nasty health risks. Choose thinner solder for delicate projects, like attaching wires to a circuit board, and thicker solder for projects involving heftier wires or bulkier pieces of metal.

SOLDERING IRON This tool has a metal tip and an insulated handle. When it's powered on, the tip heats up so it can melt solder. There are low- and high-wattage versions: Low wattage is useful for fragile projects, while high wattage is better when your projects involve bigger pieces. There are also different tips available for the soldering iron.

SOLDERING IRON STAND Buy a stand that fits your iron so you'll have a place to put it down safely when it's hot. (Leaving this thing lying around when it's turned on is a good way to burn down the toolshed before you've even made anything cool with it!)

CLIPS AND CLAMPS Soldering requires both hands, so you'll need something to hold the materials you're soldering in place. Clips, clamps, and even electrical tape can do the job.

WIRE-MODIFYING TOOLS You'll likely be soldering a lot of wire, so it's useful to have wire cutters, wire strippers, and needle-nose pliers on hand so you can manipulate the wire. Before connecting wires, you must peel back their insulation to expose the wires, so wire strippers are definitely a must.

LIQUID FLUX Soldering works best when the items being soldered are squeaky clean, so have liquid flux on hand—it chases away oxides and other goop that can make soldering difficult.

TIP CLEANER Your soldering iron's tip will get a bit nasty as you work, so keep a wet sponge on hand to periodically wipe down the tip.

HEAT-SHRINK TUBING You can use plastic heat-shrink tubing to insulate wires before you apply heat and solder them. It's available in several diameters for projects with various wire sizes.

EXHAUST FAN The fumes from soldering are not healthy to breathe, so you need good ventilation from a fan or an open window to help clear the air.

SAFETY GOGGLES Bits of hot solder can go flying as you work, so don't do it without wearing safety goggles.

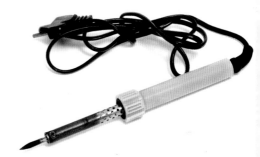

LEARN SOLDERING BASICS

Now that you have your soldering gear together, here's how to get it done.

At its most basic, soldering is simply attaching wires to wires. But when soldering onto a circuit board, the process is a little different.

SOLDERING WIRES

STEP 1 In a well-ventilated space, with your safety goggles on, plug in your soldering iron to heat it up.

STEP 2 Prepare the materials you want to join with solder. If you're connecting two wires, peel back any insulation about ½ inch (1.25 cm), and twist the wires together. Place your materials on a work surface like scrap wood.

STEP 3 Cut a length from the spool of solder and coil it up at one end, leaving a short lead. You can hold on to the coiled end as you apply the solder.

STEP 4 Touch the iron to the point where the wires are twisted together. Leave it there until the wires are hot enough to melt the solder (about 10 seconds), then touch the solder to the wire joint every few seconds until it begins to melt. Allow enough solder to melt onto the wires to cover them, then pull the solder and soldering iron away. Don't touch the solder directly to the soldering iron during this process.

STEP 5 To fix a mistake, you can desolder your joint (by reheating the solder), and reposition the components.

SOLDERING A CIRCUIT BOARD

STEP 1 Place the component that you wish to solder on the circuit board and clamp it down, then push its lead through one of the holes on the board.

STEP 2 Solder the leads to the bottom of the board. Press the soldering iron to the lead and the metal contact on the board at the point where you want them to connect. Once they heat up enough to melt the solder—just a few seconds—melt a drop of solder at the connection point.

STEP 3 Pull the solder away, then remove the soldering iron a second or two later. Once you've soldered all the leads onto the circuit board, trim off excess wire.

TRY TINNING

If you're working with components that have to be surface-mounted on a circuit board—ones that don't have leads you can thread through to the back of the board— use a technique called *tinning.*

STEP 1 Melt a drop of solder on the board where you want to attach the component. Then remove the soldering iron.

STEP 2 Pick up the component with tweezers, heat up the drop of solder on the board, and carefully place the component on the solder.

STEP 3 Hold the component in place for a few seconds until the solder cools.

STEP 4 If you need to desolder joints on a circuit board, use a desoldering pump.

003 STUDY CIRCUIT COMPONENTS

To build a circuit, first you've got to understand its parts.

Maps of how current flows through a circuit are called *schematics*. Symbols represent components, and lines show the current's path.

In this book, we use circuitry diagrams to show how to attach projects' components, so here we'll introduce you to some of the components that show up on these diagrams.

SWITCHES Switches open or close a circuit. Some are normally open as a default; others are normally closed.

TRANSISTORS A transistor amplifies energy flowing to its base pin, allowing a larger electrical current to flow between its collector and emitter pins. The two basic types of transistors, NPN and PNP, have opposite polarities: Current flows from collector to emitter in NPN transistors, and flows from emitter to collector in PNP transistors.

RESISTORS A circuit needs resistance to function. Without it, you'll end up with a short circuit, in which the current flows directly from power to ground without being used, causing your circuit to overheat and otherwise misbehave. To prevent that from happening, resistors reduce the flow of electrical current. The level of electrical resistance between two points is measured in ohms. Check to make sure a component's resistance matches the level indicated in the circuitry diagram.

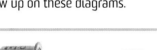

POTENTIOMETERS When you need to vary resistance within a circuit, use a potentiometer instead of a standard resistor. These have a controller that allows you to change the level of resistance: "B" potentiometers have a linear response curve, while "A" potentiometers have a logarithmic response curve.

BATTERIES These store power for a circuit, and you can use more than one to increase voltage or current.

CAPACITORS These store electricity, then release it back into the circuit when there's a drop in power. Capacitor values are measured in picofarads (pF), nanofarads (nF), and microfarads (μF). Ceramic capacitors aren't polarized, so they can be inserted into a circuit in any direction, but electrolytic capacitors are polarized and need to be inserted in a specific orientation.

INTEGRATED CIRCUITS These are tiny circuits (usually including transistors, diodes, and resistors) prepacked into a chip. Each leg of the chip will connect to a point in your larger circuit. These vary widely in their composition, and will come with a handy data sheet explaining their functions.

WIRE These single strands of metal are often used to connect the components in a circuit. Wire comes in various sizes (or gauges), and it's usually insulated.

TRANSFORMERS These devices range from thumbnail-size to house-size, and consist of coils of wire wound around a core, often a magnet. Made to transfer alternating current from one circuit to another, they can step the power of the current up or down depending on the ratio of wire windings between one coil and another.

DIODES These components are polarized to allow current to flow through them in only one direction—very useful if you need to stop the current in your circuit from flowing the wrong way. The side of a diode that connects to ground is called the *cathode*, and the side that connects to power is called the *anode*. Light-emitting diodes, or LEDs, light up when current flows through them.

PHOTOCELL A photocell is a sensor that determines how much light (or other electromagnetic radiation) is present—it then varies its resistance between its two terminals based on the amount it detects.

004 BUILD A CIRCUIT

Now you know what goes into a circuit—so make one.

STEP 1 Assemble all the components that appear on your schematic, along with any tools you'll need.

STEP 2 To test your circuit before you solder it together, set it up on a breadboard. Breadboards are boards covered in small holes that allow you to connect components without soldering.

STEP 3 Once you're ready to construct the circuit, start by installing the shortest components. This helps you avoid having to move taller components out of the way. Orient labels in the same direction so they're all legible at once.

STEP 4 Many components have lead wires that you can insert into a circuit board. Bend these leads before you insert the component.

STEP 5 You'll need to hold your parts in place while you solder the circuit together. You can do this by clinching lead wires using tape, or bracing the parts against your work surface.

STEP 6 As you solder, check that each component is aligned correctly after you solder the first pin or lead—it's easier to make adjustments at this point, before you've finished soldering a part in place.

STEP 7 When everything's soldered in place, trim all your circuit's lead wires and test it out.

005 CHOOSE A MICROCONTROLLER

To be a geek, you've got to have these microcontroller basics under your belt.

A microcontroller is essentially a tiny computer with a central processing unit (CPU), memory, and input and output. It's useful for controlling switches, LEDs, and other simple devices. Scope these features.

PROGRAMMABILITY Some can only be programmed once, while some can be erased and reprogrammed. Some allow you to add external memory.

MEMORY Microcontrollers come with a set amount of memory. Make sure that the microcontroller you choose has sufficient memory to handle your project.

COMPLEXITY For a more complex project, seek out a model with lots of input and output pins and more memory than the lower-end microcontrollers.

PHYSICAL PACKAGING A microcontroller's construction can influence how easy it is to use. For instance, less space between pins can make the device harder to work with.

PROGRAMMING LANGUAGE Different microcontrollers use differet programming languages. Choose one that uses a language you already know or are planning to learn.

SOFTWARE Some microcontrollers have easier-to-use software tools than others. If you're a beginner, ask around among your tech-savvy friends to get a sense of what's right for you.

006 PROGRAM AN ARDUINO

An Arduino is a an open-source microcontroller. Learn to program one and explore the possibilities.

STEP 1 Arduino microcontrollers come in a variety of types. The most common is the Arduino UNO, but there are specialized variations. Before you begin building, do a little research to figure out which version will be the most appropriate for your project.

STEP 2 To begin, you'll need to install the Arduino Programmer, aka the integrated development environment (IDE).

STEP 3 Connect your Arduino to the USB port of your computer. This may require a specific USB cable. Every Arduino has its own virtual serial-port address, so you'll need to reconfigure the port if you're using different Arduinos.

STEP 4 Set the board type and the serial port in the Arduino Programmer.

STEP 5 Test the microcontroller by using one of the preloaded programs, called *sketches,* in the Arduino Programmer. Open one of the example sketches, and press the upload button to load it. The Arduino should begin responding to the program: If you've set it to blink an LED light, for example, the light should now start blinking.

STEP 6 To upload new code to the Arduino, either you'll need to have access to code you can paste into the programmer, or you'll have to write it yourself, using the Arduino programming language to create your own sketch. An Arduino sketch usually has five parts: a header describing the sketch and its author; a section defining variables; a setup routine that sets the initial conditions of variables and runs preliminary code; a loop routine, which is where you add the main code that will execute repeatedly until you stop running the sketch; and a section where you can list other functions that activate during the setup and loop routines. All sketches must include the setup and loop routines.

STEP 7 Once you've uploaded the new sketch to your Arduino, disconnect it from your computer and integrate it into your project as directed.

THE FUN STUFF

007 RIG A DIY POLYGRAPH TEST

If you suspect somebody's putting you on, prove it.

COST $

TIME ⏱

EASY ●●●○○ HARD

MATERIALS

Scissors
Adhesive Velcro
Aluminum foil
Electrical wire

Wire strippers
Arduino UNO
10k-ohm resistor
USB cable

STEP 1 With scissors, cut a strip of Velcro and a strip of aluminum foil so that they are equal in length and long enough to wrap around a finger.

STEP 2 Strip the end of a piece of wire, put it on the foil's center, and place the adhesive Velcro on top to secure it, sandwiching the wire between the foil and the Velcro.

STEP 3 Flip the foil over and adhere a small piece of Velcro to this side, so that you can secure it around a finger with the aluminum foil inside—this is an electrode. Repeat this process so that you'll have two electrodes.

STEP 4 Wire the electrodes to the Arduino UNO according to the circuitry diagram at right.

STEP 5 Connect the Arduino to your computer using the USB cable. Then download the "Graph" code found at popsci.com/thebigbookofhacks onto the Arduino, and run the processing program on your computer.

STEP 6 Find a person you want answers from. Put the electrodes on one finger of each of the person's hands.

STEP 7 A graph pops up. The more the person sweats, the more conductive his skin becomes, and the higher the line of the graph goes. Skyrocketing right along with it is the likelihood that he is lying.

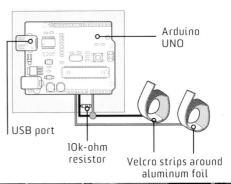

Arduino UNO

USB port

10k-ohm resistor

Velcro strips around aluminum foil

When your radio is dead, this throwback device will do the trick.

MATERIALS

Large safety pin
Toilet-paper tube
Magnet wire
Wood pencil stub
Stripped cat 6 cable
   for antenna and
   ground
Radiator or metal coat
   hanger
Wood board
Metal thumbtacks
Blued or rusty razor
   blade
Earphones

STEP 1 Use a safety pin to poke a hole in the toilet-paper tube, and secure the magnet wire to the tube by tying one end through the hole.

STEP 2 Create a coil by wrapping the magnet wire tightly around the tube 120 times, making sure that the wire is packed closely together as it coils. The number of coils affects what radio stations you'll pick up, so experiment with their arrangement if you aren't hearing your desired station clearly.

STEP 3 Make a "cat whisker" by poking the safety pin into the graphite of a pencil stub's dull end.

STEP 4 Hang the antenna cable far out the window and attach a ground cable to a metal radiator inside your home. If you don't have a radiator, attach the ground cable to a metal coat hanger and stick it in the ground outside.

STEP 5 Place the toilet-paper tube and a blued razor blade onto the wood board and push in two thumbtacks next to the tube and two next to the razor blade. Then wrap the antenna wire around the components according to the diagram below, twisting wires together when excess is needed.

STEP 6 Peel back the insulation on your headphones' audio jack. Use stripped wire to connect the safety pin in the pencil to one of the audio jack's wires, and connect the other wire to the ground.

STEP 7 Don your earphones, and touch the pencil lead to the razor blade, moving it until you start to pick up the smooth tunes you desire.

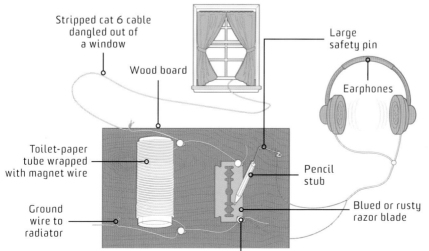

Stripped cat 6 cable dangled out of a window

Wood board

Large safety pin

Earphones

Toilet-paper tube wrapped with magnet wire

Pencil stub

Ground wire to radiator

Blued or rusty razor blade

Metal thumbtack

009 TACK UP A DIPOLE ANTENNA

STEP 1 Use wire cutters and pliers to snip a coat hanger and stretch it into a length of about 52 inches (132 cm).

STEP 2 Make a small U-shaped loop on each end of the coat hanger's wires. Then sand the loops to remove paint or coating.

STEP 3 Cut a piece of plywood to 1 inch (2.5 cm) in width and 1 foot (30 cm) in length. Screw two ½-inch (1.25-cm) sheet-metal screws about ¼ inch (6.35 mm) into the piece of wood.

STEP 4 Hook the U-shaped wire loops around the screws in the wood.

STEP 5 Slide the U-shaped tabs of a 75- to 300-ohm matching transformer under the metal screws on the plywood, one per screw. Tighten the screws until the wire ends and the transformer tabs are held together against the wood.

STEP 6 To connect a coaxial cable, screw the cable's F-type connector into the transformer.

STEP 7 Connect the coaxial cable to the radio, turn the receiver on, and position the antenna for the best signal.

STEP 8 Use a mounting bracket and screws to mount the antenna where it works the best.

010 CRAFT A CELL-PHONE "CANTENNA"

STEP 1 Use a can opener to remove the bottom of a lidless, empty can that measures 4 inches (10 cm) in diameter.

STEP 2 Solder the open end of another lidless, empty can to the first can. The total height should come to 1 foot (30 cm).

STEP 3 Solder a short piece of copper wire to an antenna connector, and drill a hole for it 3¾ inches (9.5 cm) from the closed end of the cylinder. Secure the wire in place with a nut on the inside of the cylinder.

STEP 4 Screw a passive antenna adapter cord to the antenna connector.

STEP 5 Attach the adapter cord to the back of your phone. (Don't have a smartphone? Try a basic pigtail adapter—choose one that works for your specific phone model.)

STEP 6 Enjoy improved reception. Hear what people have to say for a change.

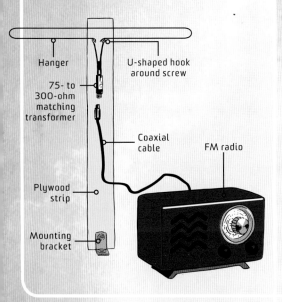

Hanger

U-shaped hook around screw

75- to 300-ohm matching transformer

Coaxial cable

FM radio

Plywood strip

Mounting bracket

Can #1

Solder

Antenna connector

Copper wire

Passive antenna adapter cord

Can #2 (with bottom still attached)

Smartphone

011 BOOST WI-FI WITH A STEAMER

STEP 1 Use tin snips to remove the steamer's center post and to make a hole about ½ inch (1.25 cm) long for the USB modem's connector.

STEP 2 Insert the USB modem into the hole with the connector end facing downward. Superglue it in place and let it dry.

STEP 3 Zip-tie two sets of two of the steamer's leaves together so that all the leaves stay open.

STEP 4 Plug the USB modem's connector end into the USB extension cable, and plug the other end of the extension cable into your laptop's ethernet portal.

STEP 5 Start picking up Wi-Fi signals from far, far away.

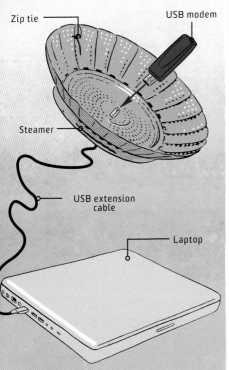

Zip tie

USB modem

Steamer

USB extension cable

Laptop

012 HANG HDTV-ANTENNA ART

STEP 1 Strip 14-gauge copper wire, then cut and bend it into eight V-shaped pieces. Place these wires onto a picture frame so that the peaks of the V shapes are 5¾ inches (14.5 cm) apart, then drill a small hole on each side of both V ends.

STEP 2 Strip 22-gauge enameled wire, then attach the V shapes to the frame by threading the enameled wire through one of the holes, crossing it over the copper wire, pulling it through the second hole, and twisting it in back.

STEP 3 Cut two lengths of copper wire to 20 inches (50 cm) in length, then bend them as shown. Lay them down the center of the picture frame so that they are 2 inches (5 cm) apart.

STEP 4 Attach the two vertical wires to the V-shaped pieces with more stripped enameled wire. Make a hole on each side of the two vertical wires' ends, then attach them to the frame as you attached the V-shaped pieces.

STEP 5 Wrap stripped enameled wire around the main vertical wires so that they nearly touch.

STEP 6 Solder two pieces of stripped enameled wire to the bottom vertical copper wires. Connect these two wires to the matching transformer, and then connect the transformer to your TV using a coaxial cable.

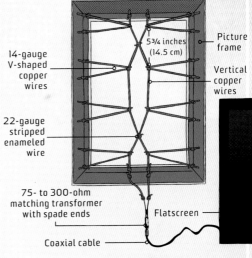

5¾ inches (14.5 cm)

Picture frame

14-gauge V-shaped copper wires

Vertical copper wires

22-gauge stripped enameled wire

75- to 300-ohm matching transformer with spade ends

Flatscreen

Coaxial cable

013 MAKE A REMOTE DISPLAY FOR YOUR COMPUTER

This stylish DIY display scrolls scores, news, weather, and anything else you tell your computer to feed it.

COST	$$
TIME	☺ ☺ ☺
EASY	● ● ● ● ○ HARD

MATERIALS

Lithium-polymer ion battery
100-mAh lithium polymer charger
Soldering iron and solder
Electrical wire
5-volt DC-to-DC step-up circuit
On/off switch
Bluetooth modem
1-by-16-character vacuum fluorescent display (VFD)
Custom case (such as a vintage radio case)
LCD Smartie software

STEP 1 Set your soldering iron to below 350°F (176°C). Solder the lithium-polymer ion battery to the charger.

STEP 2 Solder the lithium-polymer charger's output to the input of the 5-volt DC-to-DC step-up circuit. Solder an on/off switch to the step-up circuit for controlling power.

STEP 3 Connect the step-up circuit's positive and negative terminals to the Bluetooth modem and the 1-by-16-character display. Then connect the Bluetooth modem's TTL serial output to the VFD's TTL serial input.

STEP 4 Build a box to your liking, using a salvaged radio case, pieces of scrap wood, or whatever you please.

STEP 5 Use your computer's Bluetooth software for pairing and connecting the computer to the Bluetooth modem. Set the Bluetooth port for 9600, 8, N, and 1 communication parameters.

STEP 6 Download LCD Smartie. Select the test display device driver, and enter the Bluetooth COM port number and communication parameters. Set the screen size to 1-by-16-characters and apply your changes to LCD Smartie.

STEP 7 Choose the data that you want to stream and go about your business. Just stay within 50 feet (15 m) of your computer to ensure a Bluetooth wireless connection.

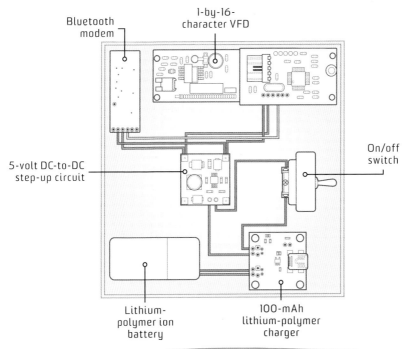

Bluetooth modem

1-by-16-character VFD

On/off switch

5-volt DC-to-DC step-up circuit

Lithium-polymer ion battery

100-mAh lithium-polymer charger

You have 1 email

Left button

Foam gasket

Right button

Left thumbstick

Face buttons

Smartphone

Home button under face button

Holes for speaker

Charger cord concealed inside here

Audio cord concealed inside here

Bottom plate

Female audio jack hidden in battery compartment

Xbox 360 controller

BUILD IT!

014 MOD AN XBOX 360 CONTROLLER INTO A SMARTPHONE CASE

Why choose between gaming on a smartphone and gaming on a console?

COST	$$$
TIME	☺ ☺ ☺
EASY ● ● ● ● ○ HARD	

If you've gotten addicted to the feel of a controller in your hands, then this smartphone case mod is for you.

MATERIALS

Xbox 360 controller
Torx screwdriver
Phillips screwdriver
Smartphone
Rotary tool
Scissors
Superglue
Foam

3.5-mm audio-jack extension cable
Smartphone wall charger with USB cable
Pliers

STEP 1 Turn over your controller and pull off the battery compartment's plastic cover. Peel away the serial-number sticker inside the battery compartment case and use the Torx screwdriver to remove the screw underneath it.

STEP 2 Remove the other screws on the controller's back. Lift the back off, and remove the bottom plate.

STEP 3 Lift out the circuit board. The thumbsticks, triggers (the buttons on the back below the left and right bumper buttons), and rumble pack motors (the cylindrical bits that dangle) will come right out with it.

STEP 4 Peel away the rubber pads that were behind the circuit board. Then turn the controller upside down and let all the buttons fall out.

STEP 5 Turn the controller back over and remove the screws on the directional pad. Use a screwdriver to release the tabs in the directional pad's center. Then remove the front and back of the pad.

STEP 6 Pull the top shelf (the upper part with the right and left bumper buttons) off the circuit board.

STEP 7 To pry the triggers from the circuit board, apply pressure to the long rod and pop up the triggers, then pinch their sides, pull out the springs, and turn them gently to pry them free. Pull the left thumbstick off the circuit board while you're at it.

STEP 8 Turn your attention to the controller. Measure a space to fit your smartphone inside it so that your phone's home key is as close as possible to a button on the controller. (This may be the green button, or it may be the left thumbstick.) Cut out this hole with a rotary tool.

STEP 9 Use a rotary tool to drill multiple small holes in the controller over your smartphone's speakers.

STEP 10 Use scissors to cut the rubber pad that was behind the face buttons, only leaving contacts for functioning buttons. (In the case of an iPhone, this is the green button.) Then reinstall the rubber pad inside the console shell with the functioning button lined up with its contact.

STEP 11 Glue all the nonfunctioning buttons back in place, along with the left thumbstick.

STEP 12 Measure and cut foam to pad the inside of the controller's upper shell, with a window the same size as your smartphone's screen. Glue the foam inside the console's hole.

STEP 13 Use a rotary tool to drill a hole in the bottom of the battery compartment. It should be big enough for the female end of your audio jack to stick out.

STEP 14 Drill a second hole in the inside wall of the battery compartment that's large enough to let you thread through the audio jack's male plug. Then pull the male end through this hole and into the main compartment.

STEP 15 Use a rotary tool to cut another hole in the battery compartment's back wall, this time large enough to fit the USB end of the smartphone's charger cable.

STEP 16 Use a rotary tool to remove the plastic casing from the smartphone's charging cable, reducing its size so it will fit inside the controller. Glue this end inside the controller's main compartment, oriented so it will plug into your phone. Then thread the USB end through to the battery compartment and coil the cord.

STEP 17 Place your smartphone inside the shell, attaching it to the charging socket and plugging the male end of the audio jack into its audio input.

STEP 18 Close the upper shell over the smartphone and secure with screws.

STEP 19 Use your phone as you normally would, and watch people do a double take.

015 Build a Hands-Free Phone

Ever wonder what people did before speakerphone? They built this gadget, which lets you hang up and still hear.

MATERIALS

Perfboard
A-4705 microphone-to grid transformer
A-3329 output transformer
Microphone
2-inch (5-cm) magnet speaker
Two CK721 transistors
0.01-µF ceramic capacitor
5-µF, 8-volt midget electrolytic capacitor
1-megohm carbon resistor

75k-ohm carbon resistor
Four AA batteries and holder
On/off switch
Electrical wire
Metal panel for electrical grounding
Wood paneling
Foam rubber
Glue
Screws
Soldering iron and solder
Saw
Screwdriver
Rotary tool

STEP 1 Assemble the electronics on the perfboard according to the circuitry diagram. Leave wire so you can later attach the on/off switch, speaker, and microphone.

STEP 2 Measure and cut wood to form the sides and bottom of a box that's large enough to fit the perfboard, then secure the box with wood glue and line the interior with foam rubber to act as a sound insulator. Glue in place.

STEP 3 Use a rotary tool to make holes on the box's sides for the speaker and on/off switch.

STEP 4 Place the assembled electronics in the box and secure the perfboard with screws. Then thread the speaker and on/off switch to the openings on the sides of the box, and screw or glue them into place.

STEP 5 Cut paneling to form the box's lid, shaped to fit your phone's handset. Cut holes for the mouthpiece and earpiece.

STEP 6 Cut and glue foam rubber so that the earpiece and mouthpiece fit the lid snugly. (These pieces will help to insulate against outside sound and feedback.)

STEP 7 Mount the lid on top of the box and glue the microphone into the opening for the earpiece under the foam rubber.

STEP 8 Place the setup near your phone. When you receive a call, place the handset on top of the box with the phone's earpiece next to the microphone. Then go about your business while Aunt Marge rattles on.

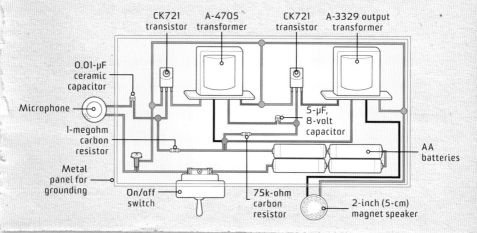

016 PLUG YOUR PHONE INTO A DINOSAUR

Popping your phone in for a charge just got a lot more Jurassic.

MATERIALS

Plastic figurine
Ruler or tape measure
Pen
Rotary tool

Smartphone (preferably with a detachable USB charger)

STEP 1 Find a plastic figurine, Mesozoic or not, that is sturdy enough to support your phone without toppling over. Make sure it has enough body mass that you can drill out a portion to fit your phone.

STEP 2 Measure how far into the body of the figurine you'll need to drill to support your phone. While you're at it, measure how wide your phone's base is. Mark on the figurine where you'll be cutting.

STEP 3 Use your rotary tool to hollow out the proper size and shape hole from your plastic figurine.

STEP 4 Once you've drilled your hole, place your phone in to test that the figurine remains stable. If it does, drill a smaller hole through the bottom of the figurine and out its underside to feed the charger through (USB-end first).

STEP 5 Plug in your charger, attach your phone, and bask in your unconventional awesomeness.

017 RIG A SMARTPHONE PROJECTOR

Beam your phone's image onto a wall for an instant big screen.

COST	$$
TIME	🕐
EASY ● ● ○ ○ ○ HARD	

MATERIALS

Narrow cardboard box
Box cutter
Fresnel lens
Audio cord
Speakers
Hot-glue gun
Modeling clay
Smartphone

STEP 1 Start with a narrow cardboard box (a shoebox does just fine) and cut a hole that's a little smaller than your Fresnel lens into one of the smaller ends.

STEP 2 Poke a smaller hole into the opposite end of the box that's big enough to allow you to run the audio cord from your phone to the speakers.

STEP 3 Using the hot-glue gun, firmly adhere the Fresnel lens over the larger hole that you made inside the box.

STEP 4 Place modeling clay on the side of your smartphone, then position it on that side inside the box with the screen facing toward the lens. (The modeling clay helps to stabilize your phone in the box, but you can still play around with its position to get a better picture.)

STEP 5 Select your entertainment of choice, then set your phone's preferences to display in landscape orientation.

STEP 6 Connect your phone to the speakers, threading the audio cord through the smaller hole in the box.

STEP 7 Close up the box, aim the lens at a blank wall, and switch off the lights. Then grab some Milk Duds and kick back with a downloaded movie or the latest episode of *The Daily Show*. Your screen should display at a totally watchable 8½ by 11 inches (22 by 28 cm).

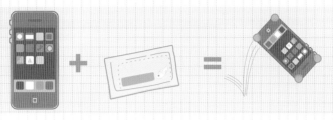

5 MINUTE PROJECT

018 MAKE A PHONE "BOUNCEABLE"

STEP 1 Break a pack of Sugru (a malleable silicone rubber that sticks to plastics) into four pieces.

STEP 2 Place one piece on each corner of your phone, folding it over onto the front and back surfaces.

STEP 3 Mold the Sugru as desired on each corner within 30 minutes.

STEP 4 Leave it to cure overnight before using, then fear dropping it no more.

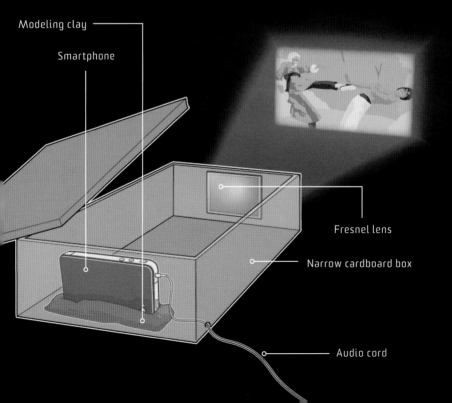

Modeling clay

Smartphone

Fresnel lens

Narrow cardboard box

Audio cord

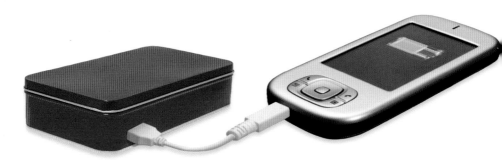

019 CHARGE A PHONE WITH SOLAR RAYS

Harness the sun's rays to keep your phone juiced on the go.

MATERIALS

Two 3-volt, 20-mA mini solar panels
Wire strippers
Small heat-shrink tubing
Heat gun
Soldering iron and solder
Cell-phone charger
Large heat-shrink tubing
Double-sided tape
Mint tin

STEP 1 Cut the wires on both mini solar panels to 1 inch (2.5 cm) in length; strip ¼ inch (6.35 mm) of the plastic coating off each wire.

STEP 2 Cut the small heat-shrink tubing into four 1-inch (2.5-cm) pieces. Cover the solar panels' two positive wires with the tubing; heat to shrink with the heat gun.

STEP 3 Solder the negative lead of one solar panel to the positive lead of the other. Cover with a piece of small tubing; heat to shrink with the heat gun.

STEP 4 Cut off 2.5 feet (75 cm) from your charger cord. Strip 2.5 inches (6.35 cm) from the loose end.

STEP 5 Cut ¼ inch (6.35 mm) off the wires inside the cord to make leads. Cover with the large tubing; heat.

STEP 6 Solder the negative leads of the charger cord wires and the solar panels together; repeat with the positive leads. Slide large tubing over them and heat.

STEP 7 Cover the backs of the solar panels with double-sided tape; then secure them to the inside of the tin.

STEP 8 Tuck the wires into the tin and close it. To use, open up the tin and let the solar panels juice up under the sun.

020 MAKE A SMARTPHONE TRIPOD

Take steady smartphone shots with a sporty improvised tripod.

Smartphone

Charger port

Tennis ball

Tripod "feet"

STEP 1 Cut a tennis ball in half.

STEP 2 With a pen, mark three "feet" on the bottom. (These will allow the tennis-ball half to balance.) There should be about 2 inches (5 cm) between each foot.

STEP 3 Cut slight arches between the tripod feet.

STEP 4 Make a slit in the top of the tennis ball and insert a charger port.

STEP 5 Plug your phone into the port and snap away.

---

5 MINUTE PROJECT

021 PROTECT YOUR TOUCHSCREEN WITH THIN VINYL

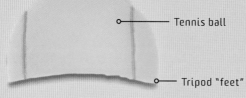

STEP 1 Measure and cut a piece of thin, nonadhesive vinyl sheeting to cover the phone's touchscreen.

STEP 2 Wipe away any dust on the vinyl and on your phone's screen.

STEP 3 Line up the vinyl with the touchscreen and slowly apply it, smoothing out air bubbles as you press it down.

022 SEW AN EASY EBOOK READER CASE

Protect your slick tablet with a felt case tucked inside an oversize book.

COST	$
TIME	☺ ☺
EASY • • ○ ○ ○ HARD	

MATERIALS

Tablet

Book 1 inch (2.5 cm) larger than your tablet on all sides

Box cutter

Felt

Scissors

Ruler

Hot-glue gun

Cardboard

Pencil

3 feet (90 cm) of ½-inch (1.25-cm) braided elastic

STEP 1 Using a box cutter, cut out the pages from the book. Glue a strip of felt to the inside of its spine.

STEP 2 Cut two pieces of cardboard to the size of the book's covers. Snip off the cardboard's corners.

STEP 3 Measure and cut two pieces of felt so they're 2 inches (5 cm) larger than the cardboard on all sides. Cut 45-degree-angle slits into the felt corners of the felt pieces.

STEP 4 Position one of the cardboard pieces in the center of a felt piece. Fold the felt over the cardboard's corners and hot-glue it in place. Then repeat with the other pieces.

STEP 5 Trace the tablet's outline on the back of one cardboard piece. On both sides of each of the outline's corners, use scissors to punch two holes large enough

to fit your elastic. Each hole should be 1 inch (2.5 cm) from the outline's corner.

STEP 6 Cut four 4-inch (10-cm) pieces of braided elastic. Feed one from the back of the cardboard up through one of the holes and then feed it back through the facing hole. Repeat with the other three strips and insert your tablet so that the strips go over the device's corners, holding it in place. If it fits, hot-glue the elastic pieces' ends to the cardboard.

STEP 7 Use scissors to cut two holes into the book's back cover near its outside edge. Thread the remaining elastic through one hole from the outside. Measure how long the elastic needs to be to encircle the book when it's closed with the tablet and cardboard tucked inside, then cut the elastic and glue its ends to the inside back cover.

STEP 8 Line up the felt-wrapped cardboard pieces with the book covers and hot-glue them together (the felt should be facing you, on the inside of the book).

STEP 9 Let the glue dry, slide the tablet under the elastic, and revel in being secretly high tech.

Felt on book's spine

Elastic strips through the cardboard

Slide and release the power switch to wake

Dick and Jan

We Play and PRETEND

Felt-wrapped cardboard

Old book with felt inside

Elastic strap

023 TURN YOUR OLD NETBOOK INTO A TOUCHSCREEN TABLET

Forget dropping big bucks on a fancy new tablet—just hack your own.

COST	$$
TIME	☺ ☺
EASY	● ● ● ○ ○ HARD

MATERIALS

Netbook
Screwdriver
Putty knife
Touchscreen overlay
Epoxy
Moldable silicone
Flash drive
Keyboard
Mouse
Retractable stylus

STEP 1 Turn off and unplug your netbook, then use a screwdriver to remove its bezel and the display's backing so that the LCD panel and its cables are exposed. Remove the keyboard and trackpad. (This may involve removing screws from the base of the netbook and prying off the top case with a flat tool, such as a putty knife.)

STEP 2 Place the netbook's exposed LCD panel over the area where the keyboard and trackpad used to be, taking care to avoid damaging the panel's cables. Be sure not to cover any areas that the netbook uses for ventilation.

STEP 3 Remove the paper on the back of the touchscreen overlay to reveal the adhesive backing. Place it over the netbook's LCD panel.

STEP 4 Plug the touchscreen overlay's USB cable into the netbook's USB port. It will be either an internal port on the motherboard, or an external port as on most computers.

STEP 5 Reattach the bezel to the front of the converted netbook with epoxy. If parts of the bezel cover the touchscreen, remove them before reattaching. If there is too much space between the bezel and the base of the netbook tablet, fill the gap with moldable silicone, sealing the two parts together. Let dry for 24 hours.

STEP 6 Copy the drivers that came with the touchscreen overlay onto a flash drive and plug the drive into the newly modified netbook tablet. Connect a keyboard and mouse to the tablet and install the drivers.

STEP 7 Run the calibration tool and then use the stylus to calibrate the touchscreen overlay.

STEP 8 Touch away on your new ad hoc touchscreen tablet, and chuckle at suckers who spent a bundle on a brand-new one.

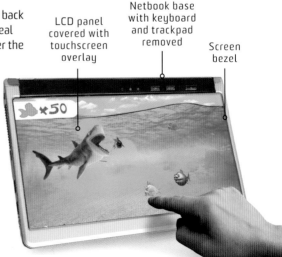

LCD panel covered with touchscreen overlay

Netbook base with keyboard and trackpad removed

Screen bezel

024 FASHION A DIY STYLUS FOR YOUR TOUCHSCREEN DEVICE

Cobble together a stylus and keep your greasy fingers off that tablet.

MATERIALS

Small scissors

Conductive foam

2-mm drafting lead holder

Plastic ink tube from a ballpoint pen

STEP 1 Use the small scissors to cut a piece of conductive foam to a cube shape about 1/4 inch (6.35 mm) in length on all sides.

STEP 2 Trim the conductive foam down further to create a rounded tip.

STEP 3 Drop the piece of foam into the lead holder and use the plastic ink tube from a ballpoint pen to push the foam down until it protrudes just out of the tip of the holder. Discard the ink tube.

STEP 4 Pinch the holder's tip to secure the foam in place. Try it out on a tablet near you.

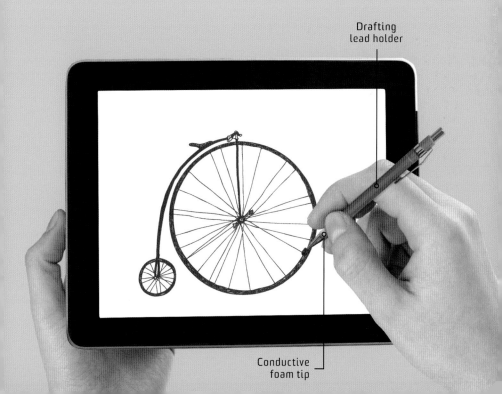

Drafting
lead holder

Conductive
foam tip

025 STASH A FLASH DRIVE IN A CASSETTE

STEP 1 Using a small screwdriver, pry off the USB drive's plastic casing.

STEP 2 Decide where you want the flash drive to poke out of the cassette. Then trace the flash drive's connector end onto that spot.

STEP 3 Remove the small screws holding the cassette together with the small screwdriver.

STEP 4 Carefully cut out the traced area using a rotary tool.

STEP 5 Wind the tape so that it's on the spool farther away from the hole for the flash drive.

STEP 6 Tape the flash drive down inside the cassette with electrical tape so that its end sticks out through the hole.

STEP 7 Reassemble the cassette, load up the flash drive with a playlist of songs, then gift it as a throwback "mix tape."

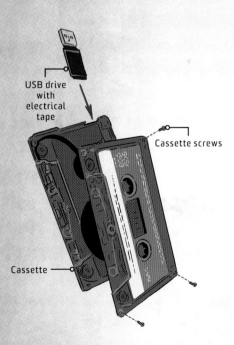

USB drive with electrical tape

Cassette screws

Cassette

026 MAKE A PINK-ERASER FLASH DRIVE

STEP 1 Remove the flash drive's plastic casing.

STEP 2 Find two erasers of the pink, parallelogram variety. Cut off the end of one, starting where the end begins to slant down. Cut the other to roughly one-third the original length.

STEP 3 Use a craft knife to hollow out both erasers. Then test to make sure the flash drive fits nicely.

STEP 4 Stick the drive inside the larger eraser, then cap it off with the smaller one. And there you have it: a discreet flash drive that can hold all of your top-secret documents—almost as if they've been "erased."

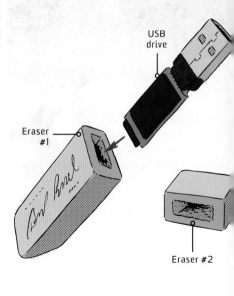

USB drive

Eraser #1

Eraser #2

027 FAKE IT WITH A SAWED-OFF FLASH DRIVE

STEP 1 Peel off the plastic cover of the USB drive. (It helps to pick a flash drive that's on the smaller side.)

STEP 2 Use a craft knife to make deep cuts in the casing along both sides of the connective end of a USB cord, piercing to the metal shell underneath. Peel off the casing to get at the inner parts.

STEP 3 Use a small screwdriver to pry apart the metal shell. Remove the "lid."

STEP 4 Underneath this lid are a few wires and miscellaneous plastic bits. They're in your way, so go ahead and cut them out with a craft knife.

STEP 5 Grab the flash drive, and protect its back (where there are metal parts that require insulation) with electrical tape.

STEP 6 Apply epoxy to the inside of the opening you've made in the end of the cord, then slide the USB drive inside.

STEP 7 Hack the cord, fray the wires as desired, and plug it into your computer. Await sounds of horror.

028 HOUSE A FLASH DRIVE IN A LEGO

STEP 1 Peel off the plastic casing on the flash drive.

STEP 2 Find a LEGO brick large enough to house the drive (a 2x6 one is ideal). Using a rotary tool, scrape out the brick's insides.

STEP 3 Measure the USB connector to get the dimensions that you'll need for the hole in the LEGO. Keep in mind that the hole should fit the USB snugly, allowing the business end to protrude and plug into your computer.

STEP 4 Draw a rectangle of these dimensions against the small end of the brick. Cut the shape out with a rotary tool.

STEP 5 Use the rotary tool to remove the top from a second LEGO of the same size and color.

STEP 6 Tape the flash drive to the top's underside with electrical tape.

STEP 7 Glue the LEGO top to the hollowed-out LEGO, allowing the USB connector to stick out the end.

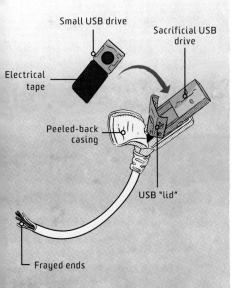

Small USB drive

Sacrificial USB drive

Electrical tape

Peeled-back casing

USB "lid"

Frayed ends

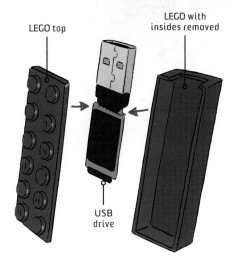

LEGO top

LEGO with insides removed

USB drive

029 HACK A FOOT-OPERATED MOUSE

COST	$$
TIME	☺ ☺
EASY ● ● ● ○ ○ HARD	

Take some strain off the old wrist with a funny foot-powered mouse.

MATERIALS

½-inch (1.25-cm) PVC sheet
Bedroom slippers
Two roller lever switches
Optical mouse
Rotary tool
Small hand file
Soldering iron and solder
Electrical wire
Metal brackets
Nuts, bolts, and washers
Screwdriver
Metal wire
Screws and nails
7 feet (2 m) of ¾-inch (1.9-cm) clear plastic tubing
Rubber doorstops

STEP 1 On the PVC sheet, position and trace your slippers the way your feet would rest while you're seated. Mark places for the left- and right-click roller lever switches near the left slipper, where your foot will operate the mouse, then outline the mouse slightly to the left of the right slipper's top.

STEP 2 Using the rotary tool, channel out spots for the roller lever switches and a hole for the mouse. Use the file to smooth the edges of the holes.

STEP 3 Remove the mouse's top cover from the base. Lift out the circuit board and remove the scroll wheel.

STEP 4 On the bottom of the circuit board, locate where the mouse's switches once connected to its buttons. Solder a length of electrical wire to each of the mouse's outboard solder connections.

STEP 5 Put the circuit board back inside the mouse cover. Then solder the lead wires from the outboard solder connections to the left and right roller lever switches—the wire from the original left-click switch to the new left switch, and the right to the right.

STEP 6 Secure two metal brackets to the mouse with two nuts, bolts, and washers. Screw the brackets into the underside of the PVC sheet so that the mouse is positioned belly-up.

STEP 7 Thread small pieces of metal wire through the mounting holes in each roller lever switch. Bend the ends and use nails to secure the switches underneath the PVC sheet.

STEP 8 Screw plastic tubing around the edges of the PVC sheet and in between where your feet go to create bumpers that help guide your feet.

STEP 9 Trim four rubber doorstops to work as risers, propping the PVC sheet up off the floor and providing clearance for the mouse. Then screw one to each corner of the footboard.

STEP 10 Plug the mouse's USB connector into your computer and begin scrolling with your feet.

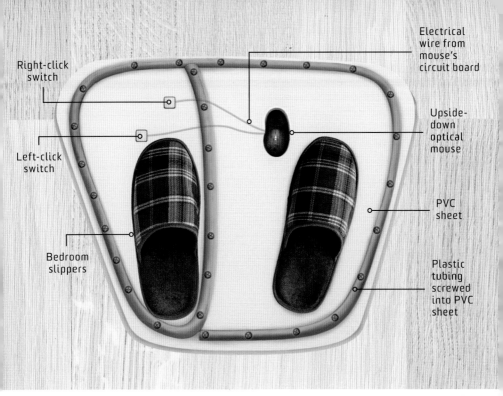

Right-click switch

Left-click switch

Bedroom slippers

Electrical wire from mouse's circuit board

Upside-down optical mouse

PVC sheet

Plastic tubing screwed into PVC sheet

030 TURN YOUR LAPTOP INTO A WHITEBOARD

STEP 1 Cut static-cling or sticky whiteboard paper to fit your laptop's top.

STEP 2 Apply the whiteboard paper to your laptop. Add an adhesive Velcro dot.

STEP 3 Stick adhesive Velcro to the back of a whiteboard marker. Then get to doodling.

031 CREATE A GLOWING MOUSEPAD

Brighten up your all-night gaming sessions with this LED-lit mousepad.

COST	$
TIME	🕐 🕐
EASY ● ● ● ○ ○ HARD	

MATERIALS

Safety goggles
Plexiglas
Tablesaw with a glass cutting blade
Rotary tool with glass-safe bit
Two small white LED lights
Electrical wire
USB connector
Clear tape
Printed design, if desired

STEP 1 Decide what size and shape you'd like your mousepad to be. Then, while wearing safety goggles, use a table saw with a glass-cutting blade to cut the Plexiglas to size.

STEP 2 Fit your rotary tool with a glass-safe bit, then use it to round the Plexiglas's edges and wear them down.

The more surface area that you make opaque, the more light your mousepad will emit.

STEP 3 Use your rotary tool to carve out a channel into the Plexiglas, starting from the top center of the glass and then forking into two channels about 1 inch (2.5 cm) down.

STEP 4 Use the rotary tool to extend the two channels parallel to the glass's top edge, ending 1 inch (2.5 cm) from the Plexiglas's edges.

STEP 5 Attach pieces of electrical wire to the LEDs' leads, then peel back the plastic on your USB cord. Next, attach the two positive wires and two negative wires on the LEDs to the USB cord's positive and negative wires.

STEP 6 Place the LEDs and wires into the carved channels and secure them with clear tape. Cover the mousepad with a design, if you like.

STEP 7 Plug the USB into your computer, dim the lights, and get your game on.

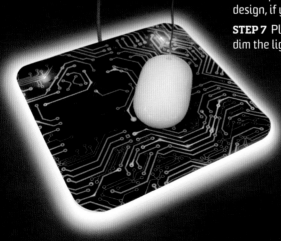

032 RIG A SUPERPORTABLE KEYBOARD

Cut a keyboard to get to the touch-sensitive membrane inside.

MATERIALS

USB keyboard
Screwdriver
Transparent contact
  paper

Superglue
Adhesive stickers

STEP 1 Use a screwdriver to deconstruct the keyboard. The good stuff is in the middle: Specifically, the three-layer membrane and the attached control board, which feeds the USB wire.

STEP 2 Remove and reserve the membrane's control board and switch pad (the rubber pad that presses the control board's contacts to the membrane).

STEP 3 Using the switch pad, board, and nuts and bolts, reassemble the membrane to the control board. The traces of the membrane should line up with the traces on the control board. (If your keyboard had a socket and ribbon cable, reinsert the cable.)

STEP 4 Cover both sides with transparent contact paper, and apply glue to the edges to keep the three membranes in place.

STEP 5 Apply adhesive stickers for each key, taking care to place the stickers on the keys' contacts. (For instance, the space bar is huge, but the contact is small, so you'll want to be sure that you put the sticker directly on the contact—not just on the key.)

STEP 6 Roll it up and be ready to type anywhere.

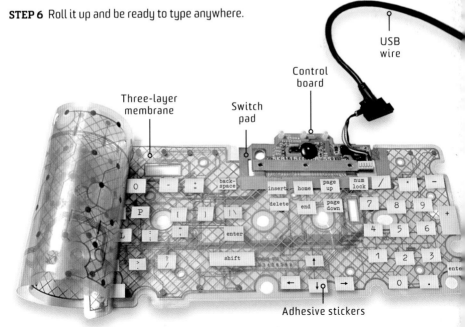

USB wire

Control board

Three-layer membrane

Switch pad

Adhesive stickers

033 TURN A PRINTER INTO A DOCUMENT SHREDDER

Got mail? Shred it and other sensitive documents with this hacked printer.

COST $$$

TIME ☺ ☺ ☺

EASY ● ● ● ● ● HARD

MATERIALS

Laser printer that prints 25 pages per minute
Screwdriver
Small Perfboard panel
Two LM3903 integrated circuits
Two 100μF, 6-volt electrolytic capacitors
Electrical wire
Wire strippers
Soldering iron and solder

6-volt, 60-mA DC motor
Drill
Glue
5-volt DC-power supply
Cheap, lightweight paper shredder (the kind that fits over a wastebasket)
Wire nuts

STEP 1 Unplug and turn off your printer, then use a screwdriver to remove its rear-access panel. Take out the fuser, which bonds toner to paper and is a fire hazard.

STEP 2 Remove the printer's top and side panels and gut it: Take out any power supplies, computer processing units, motors, fans, laser units, toner cartridges, and related circuit boards and wiring. Leave in any solenoids (the long wire loops that control the printer trays) and the toner cartridge, which will control the drive for the trays. Reattach the top and side panels.

STEP 3 To make the controller, mount the two LM3903 integrated circuits

onto Perfboard. Attach each capacitor to one integrated circuit, soldering the capacitors' positive leads to the integrated circuits' 2 pins, and their negative ones to the 1 pins.

STEP 4 Mount the controller in the case, then solder each of the printer's original solenoids to one of the two integrated circuits: The solenoids' negative leads go to the 8 pins. Their positive leads should go to the 6 pins.

STEP 5 Replace the printer's original motor with the 6-volt, 60-mA DC motor. Mount it inside the printer with the original motor's screws.

STEP 6 Take the 5-volt DC-power supply and glue it inside the case on a side panel near the original power supply's location. Then strip its output cord and connect its positive and negative wires with the positive and negative wires of the power leads on the DC motor and the shredder. (The shredder's wires should run through the back of the printer at this point.)

STEP 7 Attach wires to the 5-volt DC-power supply's positive and negative leads, then connect them to the integrated circuits: The positive leads go to the 5 pin. The negative leads go to the 4 pin. Secure all the positive and negative connections with wire nuts.

STEP 8 Replace the side panel to cover the motor. (Cut a hole for clearance if parts stick out, but cover those later with panels to avoid injury.) Drill holes in the printer case's back, then secure the shredder to it with screws.

STEP 9 Load your trays with shreddables, and start making confetti out of your private information.

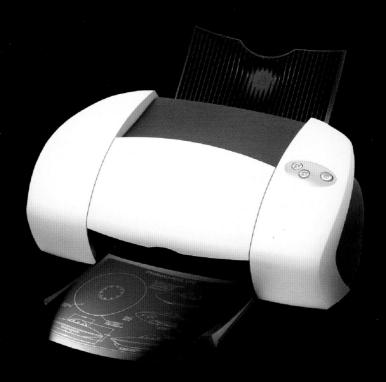

034 PRINT SECRETS WITH INVISIBLE INK

Bet the CIA sure wished it had come up with this printer hack.

MATERIALS

Inkjet printer   Syringe
UV invisible ink   UV lamp

STEP 1 Open up your inkjet printer and extract one of the ink cartridges. Remove its cap and pull out the sponge.

STEP 2 Rinse out the sponge and the inside of the cartridge until the water runs clear, then put the clean sponge back inside the cartridge.

STEP 3 Use a syringe to inject invisible ink into the sponge. Replace the cartridge's cap and put the cartridge back inside the printer.

STEP 4 Adjust your computer's settings to print using only the color that you've replaced with invisible ink.

STEP 5 View the secret info on the documents by holding them up to a UV lamp.

035 TRICK OUT YOUR COMPUTER TOWER WITH ENGRAVING

This is one tower mod that doesn't belong hidden under your desk.

MATERIALS

Paper and pencil or
  printer
Computer tower
Screwdriver
Spray paint
Plexiglas sheet
Masking tape

Safety goggles
Rotary tool
Small engraving bits
Screws and bolts
Fluorescent strip, if
  needed

STEP 1 Draw or print your design on a sheet of paper, sized to fit your tower's side panel.

STEP 2 Unscrew the metal side panel from your computer tower and set it aside for later.

STEP 3 Spray-paint a sheet of Plexiglas with a color to match your case so only the engraved design will be visible later. Let dry.

STEP 4 Tape the drawn or printed design onto your Plexiglas sheet, being sure to tape it over the painted side.

STEP 5 Put on your safety goggles and, using the rotary tool, begin engraving. Follow the lines of your design on the paper and etch it on the Plexiglas beneath. Work carefully and slowly to avoid ruining your design.

STEP 6 Remove the taped-on template and wipe down the engraved Plexiglas.

STEP 7 Measure and cut a hole that's slightly smaller than your Plexiglas in the metal panel.

STEP 8 Drill four holes into the corners of the Plexiglas and the case. Attach the etched panel to the metal frame with screws small enough that it will fit inside the case.

STEP 9 If your machine lacks a fluorescent strip, install one, either plugging it into the power supply or directly to the circuit board. (Which you choose depends on the bulb you've purchased, and what sort of machine you have.)

STEP 10 Use the original bolts and screws to reattach the metal panel with the Plexiglas inside the tower.

036 TURN ON YOUR COMPUTER WITH A MAGNET SWITCH

Dupe would-be information thieves with a handy on/off switch mod.

MATERIALS

Screwdriver
Computer tower with plastic front panel
Wire strippers
Reed switch
Steel or iron nut
Electrical tape
Small magnet

STEP 1 Open your computer tower and remove the front panel, exposing the wires attached to the power button. Cut one of the wires, and strip both ends.

STEP 2 Place the reed switch between the two wire ends and twist them together, sandwiching the reed switch.

STEP 3 Tape the reed switch to the inside of the front panel of your computer and tape or glue the steel or iron nut next to it. Close the computer back up.

STEP 4 To turn on your computer, push the power button and stick a magnet to the case where the reed switch is located—the nut should hold the magnet in place. If someone tries to turn your computer on without the magnet, they'll have no luck.

5 MINUTE PROJECT

037 MAKE AN EXTERNAL HARD DRIVE

STEP 1 Salvage a working hard drive from a laptop or a computer tower.

STEP 2 Locate the ports on a hard drive case's baseplate, then attach them to the hard drive.

STEP 3 Line up the holes in the drive and baseplate and screw them together.

STEP 4 Slide these parts inside the case. Screw on the faceplate, lining up its holes with the ports to keep them accessible.

038 DYE YOUR LAPTOP

Fight the monotony of boring case colors with a laptop dye job.

MATERIALS

White plastic laptop
Small screwdriver
Sandpaper
Denatured alcohol
Paper towel
Rubber gloves

Deep pan
8 cups (2 L) water
Fabric dye
2 tablespoons table salt

STEP 1 Carefully take apart your laptop. You'll likely need a small screwdriver to remove the screws that hold the case, battery, and other parts in place. Separate the plastic parts (the ones you'll be dyeing) from the metal and electronic parts.

STEP 2 Sand down the plastic pieces to remove the glossy layer, which prevents dye from absorbing quickly and evenly. Leave it if you don't want a matte laptop, but the process takes longer and results in splotchiness.

STEP 3 Clean all the parts with denatured alcohol and a paper towel, then let them dry.

STEP 4 Wearing rubber gloves, fill your deep pan with 8 cups (2 L) of water. Add the dye to the water along with 2 tablespoons of table salt. Stir together.

STEP 5 Place the pan on the stove and heat. When the water starts to boil, submerge the part you wish to dye.

STEP 6 Add water if the liquid boils off, and stir every once in a while. Larger parts require more time.

STEP 7 Once you're satisfied with a part's dye distribution, remove it from the bath, wipe it down, and rinse it with cold water. Dry thoroughly.

STEP 8 Reassemble your laptop. Resume being awesome.

039 MAKE A STEAMPUNK-INSPIRED LAPTOP CASE

Give your high-tech machine an old-school Victorian vibe.

COST	$
TIME	🙂 🙂 🙂
EASY	● ● ● ○ ○ HARD

MATERIALS

Laptop
Tracing paper
Double-sided heavy duty adhesive sheet
Wood veneer sheeting
Sandpaper
Craft knife
Hot water
Masking tape
Paint and paintbrushes
Polyurethane
Superglue
Miscellaneous embellishments

STEP 1 Lay tracing paper over the back of your laptop and trace where the wood veneer will go, leaving holes for plugs, fans, and hatches that allow you access to the computer's insides.

STEP 2 Open up your laptop and, using the same method, create patterns for the frame around the screen and the area surrounding the keyboard.

STEP 3 Apply a heavy-duty adhesive sheet to the blank side of the wood veneer sheeting.

STEP 4 Sand the veneer for a smooth look and feel.

STEP 5 Tape the tracing paper onto the wood veneer and cut it out using a craft knife. Cut out the veneer pieces to go around the screen and keyboard, too.

STEP 6 In the corners of the veneer, cut diagonal slits so that you can fold the veneer over the laptop's corners.

STEP 7 Soak the veneer in hot water to make it pliant. Be sure to dry off excess water before applying the veneer—for extra security, remove the battery before placing the veneer on the laptop.

STEP 8 Use masking tape to secure the wood to the laptop. Let it dry so it can mold nicely to the computer's shape, then remove it and set it aside.

STEP 9 On your laptop, use tape to mask off any areas that the veneer won't cover, including the hinge. Then paint those areas a color of your choosing.

STEP 10 Peel away the veneer's backing and apply the veneer to the laptop, starting with the inside pieces. Don't press down until you've got it lined up perfectly, then slowly press out air bubbles as you apply it. Cut away excess.

STEP 11 Mask off areas that aren't covered with veneer, then coat the veneer with polyurethane.

STEP 12 Add any desired embellishments with superglue. Apply another coat of polyurethane to make it for keeps.

040 MAKE YOUR KEYBOARD GLOW IN THE DARK

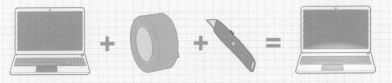

STEP 1 Carefully clean your keyboard using rubbing alcohol and a cotton swab.

STEP 2 Use a craft knife to cut squares that fit your keyboard's keys out of glow-in-the-dark tape. Slice off the squares' corners.

STEP 3 Align each square of tape you cut out over a key before pressing and smoothing it down.

STEP 4 Turn out the lights and keep on typing with ease.

Painted hinge

Wood veneer sheeting

Steampunk touches

041 SET UP A LAPTOP COOLING SYSTEM

Chill down your machine with copper's thermal conductivity.

COST	$$
TIME	☺ ☺
EASY ● ● ● ● ○ HARD	

MATERIALS

Rotary tool
Paper
Scissors
Sheet of 0.5-mm copper plating
Tin snips
6-mm center-tapped lip and spur drill
Two pieces of 6-mm copper tubing, each 2 inches (5 cm) in length
Solder and soldering iron
Plastic tubing
Rubber tubing
Bilge pump

STEP 1 Use the rotary tool to cut away at your laptop's plastic casing until the fins of your computer's internal radiator and heat sink are exposed.

STEP 2 Experiment to determine how big the fins of your heat extractor should be. Try inserting strips of paper between the fins of your computer's heat sink, cutting them down until they fit perfectly. For maximum cooling, the fins should fit as deep into the heat sink as possible.

STEP 3 Once you've determined the necessary measurements for your copper heat extractor's fins, clean the copper sheet with soap and water. Using one of your paper fins as a template, trace seven copper fins.

STEP 4 Trace two holes 6-mm in diameter along one of the short edges in each fin, positioning the holes 5/8 inch

(16 mm) apart. The 6-mm copper tubing should fit snugly through these holes, once you've drilled them.

STEP 5 Cut out the copper fins with tin snips and use the lip and spur drill to cut out the holes. Set the drill to a slow speed and do your drilling on a flat surface to prevent the copper sheet from warping.

STEP 6 Thread the fins onto the two lengths of copper tubing. You can temporarily place coins between the fins to help space them evenly so that they'll line up with the heat sink's indentations. Solder the tubes and fins together.

STEP 7 Cut a piece of plastic tubing to a length of about 1½ inches (3.75 mm). Heat it until you can bend it to fit over both pieces of copper tubing on one side of the fin-and-tubing contraption. Then insert the contraption into your computer's heat sink.

STEP 8 Hook two lengths of rubber tubing to the bare pieces of copper tubing that are plugged into your computer's heat sink. Connect these two lengths of rubber tubing to a bilge pump filled with water. Power it up and the water will circulate, carrying heat from the heat sink and keeping your computer cool.

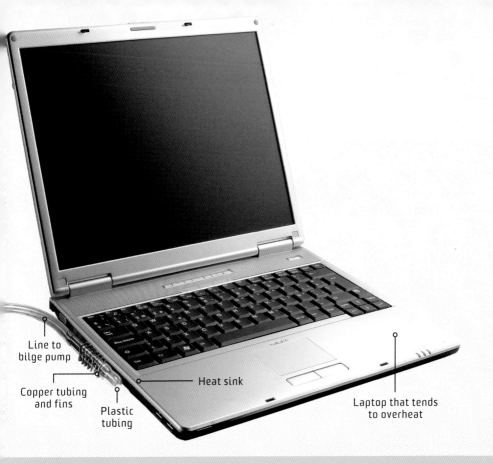

Line to
bilge pump

Copper tubing
and fins

Plastic
tubing

Heat sink

Laptop that tends
to overheat

042 CONTROL YOUR MOUSE FROM AFAR

Direct your mouse with a simple laser pointer.

MATERIALS
Optical mouse
Laser pointer, less than
  10 mW

STEP 1 Lean your mouse against your computer monitor so that it's propped upright with its belly facing out.

STEP 2 Identify your mouse's sensor, which probably looks like a tiny bubble tinted black.

STEP 3 Shine the laser pointer directly at the sensor. Once you have a lock on it, you can move it around in the mouse's vicinity to control your computer's cursor.

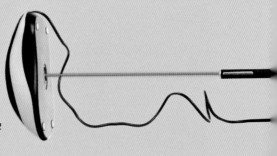

043 FAKE OUT THIEVES WITH A DESKTOP HACK

Deter computer raiders with a wallpaper that looks just like your desktop—but isn't.

STEP 1 Take a screenshot of your computer's desktop, then set the screenshot as your wallpaper.

STEP 2 Hide the real icons inside another folder.

STEP 3 Sit back and watch people try to open your desktop's unclickable folders.

Private

Secrets

Confidential

More secrets

Don't look!

Embarrassing

---

5 MINUTE PROJECT

044 ADD KEYBOARD THUMBTACKS

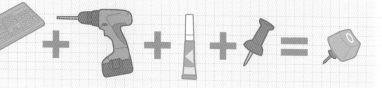

STEP 1 Remove the keys from the keyboard, and cut off the excess plastic on the back of each key with a rotary tool.

STEP 2 Widen the hole in the back of each key using a rotary tool. Put a dab of glue in the hole.

STEP 3 Insert a pushpin into the hole, pointy end facing out. Pin up something important.

045

SHIELD YOUR SCREEN FROM PRYING EYES

Sick of snoops looking at your screen? Improvise your own privacy monitor.

MATERIALS

LCD monitor
Craft knife
Paintbrush
Paint thinner
Paper towels

Piece of plastic
Old glasses
Tape
Superglue

STEP 1 Unplug an old LCD monitor and remove the plastic frame around it.

STEP 2 Use a craft knife to cut around the screen's edge, then peel back both the polarized and the antiglare films. Hang on to the polarized layer and remember its orientation.

STEP 3 Apply paint thinner to loosen the glue on the monitor's screen—don't drip it on the monitor's frame. Then wipe it off with paper towels and scrape off the softened glue with a piece of plastic.

STEP 4 Reassemble your monitor. At this point, when you turn it on, the screen looks white and blank. If you hold up the polarized film, you should see images on the screen.

STEP 5 Pop the lenses out of a pair of old glasses. Tape the lenses to the polarized film and trace around them.

STEP 6 Hold the polarized film lenses up as though you were wearing them, and look at the monitor. If you can see the images on the screen, cut out the lens-shaped film pieces with your craft knife.

STEP 7 Glue the polarized lenses onto your glasses. Put them on, and enjoy the invisible images on your screen.

046 UPCYCLE AN OLD CIRCUIT BOARD

So a gadget's circuit board is down for the count. There are still countless things you can do with it.

AN EVEN GEEKIER CLIPBOARD
Use a heat gun and pliers to strip off all the solder and bits and bobs, then apply laminate to make it smooth. Swap the clip from an old clipboard onto your new, high-tech version.

NERDTASTIC GUITAR PICK
Use a soldering iron to remove any electrical components on the circuit board, then use a rotary tool to cut out a guitar pick shape. Sand it until it's smooth and start picking.

META MOUSE PAD
Desolder a circuit board so that it's bare and cover both sides with vinyl. Plop your new mousepad on your desk and get your scroll on.

LIGHT UP THE CIRCUIT
Form a box shape with four stripped circuit boards and drill holes in their corners. Fasten them together with zip ties, and hook this box up to a hanging light-socket assembly for some nice spotlight action.

047 MAKE A LAPTOP STAND FROM A BINDER

STEP 1 Using a metal saw, cut a piece of aluminum rail so it's the length of a ring binder. Then use a metal file to round the rail's edges so you don't get scraped.

STEP 2 Place double-sided tape on the inner side of the aluminum rail.

STEP 3 Drill two sets of two holes big enough for bolts—one set through the rail and the other set through the binder.

STEP 4 Line up the holes and attach the rail and binder with the bolts, securing them with a nut on the underside. Cover these bolts with tape to avoid scratches on the laptop.

STEP 5 Measure and cut a strip of no-skid felt to the dimensions of the rail, then adhere it to the inner side of the rail. Measure and cut a larger sheet of felt so that it covers the top of the binder and secure it with adhesive. It will prevent your laptop from sliding around.

STEP 6 Use a rotary tool to cut a hole into the binder's corner for cords to pass through.

048 BUILD A USB HUB INTO YOUR DESK

STEP 1 Remove your computer, cords, and other electronics from your work station to protect them from sawdust.

STEP 2 Measure your hub and mark a spot for it on a wood desktop using a pencil and ruler.

STEP 3 Use a jigsaw to cut out the opening, tracing your marked lines. It's better to cut it slightly smaller than your hub (you can always sand it) than too big (you'd have to fill any gaps with caulking).

STEP 4 Remove the plug of wood and insert your USB hub to make sure it fits. Use sandpaper to adjust and smooth the opening.

STEP 5 Use epoxy to attach the USB hub to the inside of the hole; let dry.

STEP 6 Run the input cord from the USB hub up behind your desk to your computer.

STEP 7 Next time you need to plug something in, forget reaching around your monitor: Just plug it into the hub on your desk.

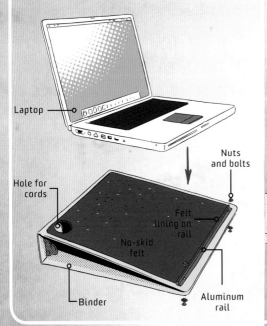

Laptop

Hole for cords

Nuts and bolts

Felt lining on rail

No-skid felt

Binder

Aluminum rail

Computer

Input cord

USB hub

Wood desk

Gadget in need of syncing

049 STASH YOUR PRINTER IN A DRAWER

STEP 1 Remove the front panel of a drawer and drill a hole into the back panel for cables. (It's better to do this in a lower drawer so that the weight of your printer doesn't stress the structure of the furniture.)

STEP 2 Measure and cut two evenly spaced recessed areas for the utility hinges on the front edge of the bottom panel. This way, the bottom and front panels will line up neatly when you reattach the front.

STEP 3 Line up the front and bottom panels and screw on the utility hinges.

STEP 4 With the front panel lowered, screw the support hinges to the drawer's side panels and then to the front panel.

STEP 5 Place your printer in the drawer and feed the cables through the hole in the back to attach them to your computer (or, if it's wireless, to a power source).

050 MOUNT STUFF BEHIND YOUR MONITOR

STEP 1 If you have limited desk space, use the back of your monitor for storage. Apply double-sided tape or adhesive Velcro strips to your computer monitor's back, or purchase plastic hooks with suction cups or Velcro backing.

STEP 2 Begin attaching office supplies you need—tape, tissues, notecards, a stapler, or a holder for pens and scissors—and maybe some you don't. (Candy and headphones can make the workday go faster.)

STEP 3 Be sure to use only lightweight materials so that your monitor doesn't tip over. Also, keep the fan clear of obstacles. Otherwise, your machine could overheat, which is much more inconvenient than not having any drawers.

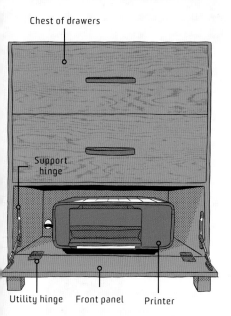

Chest of drawers

Support hinge

Utility hinge   Front panel   Printer

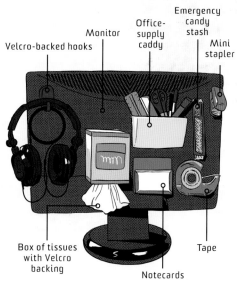

Emergency candy stash

Office-supply caddy

Monitor

Mini stapler

Velcro-backed hooks

Box of tissues with Velcro backing

Notecards

Tape

051 GET PUMPED WITH A CD DUMBBELL

CDs are pretty much obsolete. But having jacked arms never gets old.

MATERIALS

½-inch (1.25-cm) solid threaded rod

Ruler

Permanent marker

Table vise

Reciprocating saw or hacksaw

Four ½-inch (1.25-cm) washers and nuts

About 150 CDs

½-inch (1.25-cm) wrench

STEP 1 Measure and mark 6 inches (15 cm) from one end of the rod. (This is where the first CD stack will end.)

STEP 2 Place one hand on the rod, leaving about ½ inch (1.25 cm) of clearance between your hand and the mark. Make a second mark that's about ½ inch (1.25 cm) from the opposite side of your hand.

STEP 3 Measure and make a third mark 6 inches (15 cm) from the second mark for the second CD stack.

STEP 4 Place the rod in a table vise. Saw off any excess at the third mark with a reciprocating saw or a hacksaw.

STEP 5 Thread a nut onto both sides of the rod to the marked lines in the center. Add a washer on both sides.

STEP 6 Put about 75 CDs on each end, and slide on a washer and nut on both ends. Tighten with a wrench.

STEP 7 Pop your new 10-pound (4.5-kg) dumbbell out of the vise and do a few reps—you might not get ripped, but you're well on your way to getting sculpted biceps.

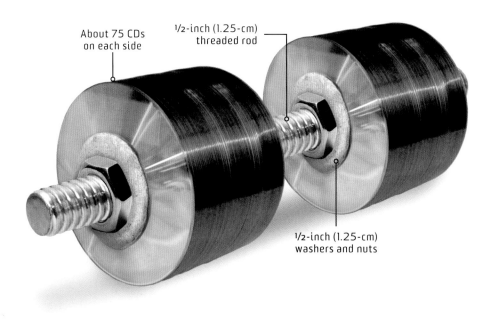

About 75 CDs on each side

½-inch (1.25-cm) threaded rod

½-inch (1.25-cm) washers and nuts

052 MAKE A FLOPPY-DISK BOX

Solve the double problem of a messy desk and a surplus of useless floppies with one simple DIY craft.

MATERIALS

Five floppy disks
Drill
Twelve zip ties
Scissors

STEP 1 Locate the tiny dimples on the back of each floppy disk. Then drill through these dimples, repeating until you've made holes in each corner of four of your five disks.

STEP 2 Place two of the four floppies in front of you so that their holes are aligned. Thread a zip tie through the two holes at the bottom, and then another zip tie through the two holes at the top.

STEP 3 Repeat with two more disks, and then connect these four floppies into a box shape.

STEP 4 Drill four holes in the fifth floppy, this time slightly above the dimples. This is the box's bottom.

STEP 5 To secure the bottom floppy to the box, line it up with one of the sides at a 90-degree angle. Thread a zip tie through the bottom's holes and the holes in the side.

STEP 6 Close the box and thread zip ties through the remaining holes in the bottom and side pieces. Cut off the ends of the zip ties.

STEP 7 Tighten and trim the zip ties, then stock the box with pens and revel in your newfound tidiness.

053 ORGANIZE LOOSE CABLES

STEP 1 Remove the dome from a CD spool. Cut a slit on the bottom of the dome; cut another slit opposite it.

STEP 2 Wind cables around the spool and run the ends through the slits.

STEP 3 Tug on one end of a cable to adjust how much of it extends outside the dome.

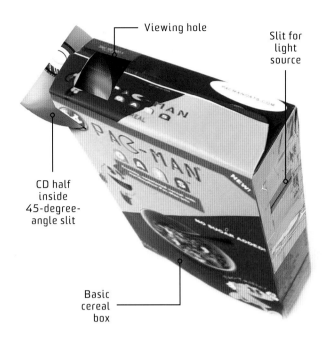

Viewing hole

Slit for light source

CD half inside 45-degree-angle slit

Basic cereal box

054 ASSEMBLE A CEREAL-BOX SPECTROMETER

See the rainbow inside everyday light sources with this easy setup.

MATERIALS

Safety glasses
Thick gloves
CD
C-clamp

Craft knife
Cereal box
Tape
Light source

STEP 1 Wearing a pair of safety glasses and thick gloves, clamp a CD down on a surface edge. Score it across the center with the craft knife, then break it in half.

STEP 2 Make a horizontal slit about 1 inch (2.5 cm) in length in one side of a cereal box, near the box's top. It should be about the width of a coin.

STEP 3 In the opposite side of the box, straight across from the first incision, make another slit. Then extend this cut to the front and back of the box, sloping down at a 45-degree angle with your craft knife. It should be deep enough for the CD half to at least partially slide into. Secure the CD in place with tape.

STEP 4 Cut a 1/2-by-1/2-inch (1.25-cm-by-1.25-cm) square hole in the box above the CD slice.

STEP 5 Hold the box up with the slit facing your light source. Look through the viewing hole at the top to see the rays of light separated on the CD inside the box.

055 RIG A SUPERSIMPLE RADIATION DETECTOR

Fear fallout no more with a device that tells you when radiation levels are high.

COST	$$
TIME	◔
EASY	● ● ○ ○ ○ HARD

MATERIALS

Aluminum can
Craft knife
NPN Darlington transistor
Electrical tape
Electrical wire
9-volt battery and snap
4.7k-ohm resistor
Soldering iron and solder
Aluminum foil
Rubber band
Multimeter

STEP 1 Make a hole in the can with a craft knife. Bend the transistor's base leg down into the can and tape it in place.

STEP 2 Use electrical wire to attach the transistor's collector leg to the negative pole of the battery snap. Keep the transitor's leads sticking up, away from the can.

STEP 3 Solder one lead of the resistor to the can's base, near the edge. Attach the other lead to the positive pole of the battery snap with a piece of wire.

STEP 4 Tape the battery snap to the side of the can and then hook the battery up to the snap.

STEP 5 Cover the can's open end with aluminum foil. Pull the foil taut, then secure it with a rubber band.

STEP 6 Attach one probe of the multimeter to the transistor's emitter leg, and the other to the wire between the resistor and the battery.

STEP 7 Turn on the multimeter and wait for the reading to stabilize—avoid touching the can or moving around near it. Once it stabilizes, you'll have the baseline reading for the radiation in the room. Keep the can away from power sources, which could confuse its reading.

STEP 8 To measure the radioactivity of an object relative to the baseline, simply place it beside the can's end and observe the changed reading on your multimeter.

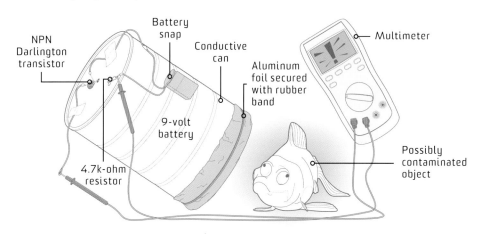

NPN Darlington transistor
Battery snap
Conductive can
Multimeter
Aluminum foil secured with rubber band
9-volt battery
4.7k-ohm resistor
Possibly contaminated object

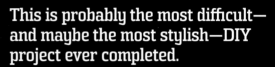

056 A HOMEMADE SCANNING ELECTRON MICROSCOPE

This is probably the most difficult— and maybe the most stylish—DIY project ever completed.

Ben Krasnow has built his share of odd contraptions, including a liquid nitrogen generator made from an air conditioner. Now, wanting a real challenge, he decided to try his hand at the toughest project he could imagine: a homemade scanning electron microscope, or SEM, fashioned from an old oscilloscope, a glass bell jar, and a refrigerator magnet. "I wanted to see if it was possible," he says.

Krasnow first spent a few weeks teaching himself the complex physics behind the instrument. Next he trolled the Internet for cheap components, sorted through his home shop for power sources that might work, and, finally built what he couldn't find.

He made his electron gun out of a thin tungsten wire, which, when heated, releases clouds of electrons that speed through a thin copper pipe toward a sample. The electron beam is then focused by the magnet. His completed microscope delivers about 50x magnification—a far cry from commercial SEMs' 1,000x or more—but experts say that doesn't lessen the accomplishment. William Beaty, a research engineer who had hoped to build the first DIY SEM, put it simply: "D'oh!"

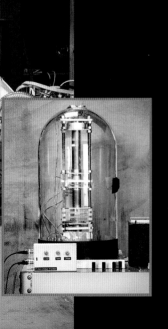

RETRO STYLE
Krasnow carefully sourced components— such as the glass bell jar and toggle switches—that would create a 1960s-space mission aesthetic.

BUILD IT!

057 SET UP A PLASMA GLOBE INSIDE A SIMPLE LIGHTBULB

Make a miniature Tesla coil inside this ubiquitous household fixture.

COST	$$
TIME	⊙ ⊙
EASY ● ● ● ● ● HARD	

MATERIALS

Ferrite-core flyback transformer (salvaged from an old CRT television)
Wire strippers
Screwdriver
Hacksaw
18-gauge wire
22-gauge wire
Electrical tape
Rotary tool
Project box
60-watt or higher clear lightbulb
24-volt DC-power supply
On/off switch
Superglue
Electrical wire
Lightbulb socket
2N3055 transistor and heat sink
5-watt, 27k-ohm resistor
5-watt, 240k-ohm resistor
Soldering iron and solder

STEP 1 Salvage a transformer from an old CRT television. To do this, turn off the television and unplug it, then open up its back and locate the transformer. (It's the bulky square metal ring with two cylindrical "cores" on it, bolted into the television's circuit board.)

STEP 2 Remove the transformer by clipping away the wires (leaving extra wire on the transformer itself) and unscrewing the bolts holding it to the circuit board.

STEP 3 There are two sets of windings on the transformer's primary and secondary core, both encased in plastic. Use a hacksaw to cut through the plastic around the windings on the smaller primary core. Then cut through and remove the windings.

STEP 4 Wind 18-gauge enameled wire around the spot where the windings used to be about five times. Then wrap 22-gauge wire four times next to the 18-gauge windings. Secure both wires with electrical tape.

STEP 5 Use a rotary tool to make three holes in a project box: one for your lightbulb socket, one for the power supply, and one for the on/off switch. Super-glue these parts in place on the top and side of the box. Place the transformer inside the box.

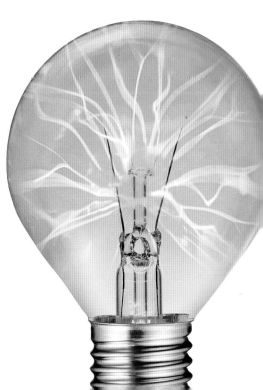

STEP 6 On the secondary core, there is another set of windings. Pry the two loose ends off the core and twist them to the connectors on the lightbulb socket.

STEP 7 Take one end of the 18-gauge wire wrapped around the transformer and connect it to the transistor's case post, and connect the other end to the power switch.

STEP 8 Take one end of the 22-gauge wire and attach it to the transistor's base post. Attach the 22-gauge wire's other end to a wire between the 27k-ohm resistor and the 240k-ohm resistor.

STEP 9 Complete the rest of the circuit as shown below using electrical wire, then screw your lightbulb into the socket, plug the contraption into your power supply, and turn on the switch.

STEP 10 Kill the lights and watch your plasma globe glow.

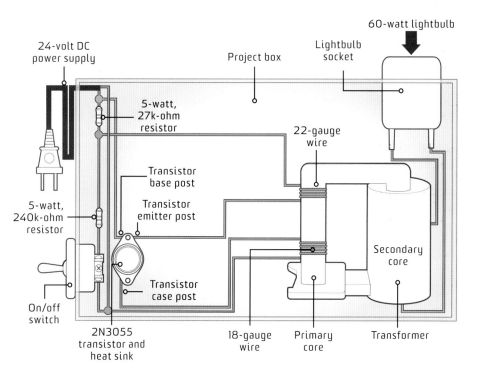

24-volt DC power supply

5-watt, 27k-ohm resistor

5-watt, 240k-ohm resistor

Transistor base post

Transistor emitter post

Transistor case post

On/off switch

2N3055 transistor and heat sink

Project box

22-gauge wire

18-gauge wire

Primary core

Secondary core

Transformer

Lightbulb socket

60-watt lightbulb

WARNING
Tesla may have invented alternating current, but he sure didn't like to see people get electrocuted with it. Avoid this fate by exercising extreme caution when handling the transformer, as its high voltage is immensely dangerous.

058 DECIMATE STUFF WITH A DIY LASER CUTTER

It used to be lasers were just for scientists. Now they're for anyone with an outdated PC.

COST	$$
TIME	☺☺☺
EASY ●●●●○ HARD	

MATERIALS

Old computer with a DVD burner
Soldering iron and solder
Knife
Vise
Metal file
Pliers
Tweezers
AixiZ module
LM317 regulator
3k-ohm resistor
Wire strippers
Drill
Thermal glue
Laser safety goggles

STEP 1 Unplug the computer and open its case. Locate and remove the power supply, heat sinks, and DVD burner.

STEP 2 Open the DVD burner, lift the circuit board, and remove the sled underneath (the part with the laser diode).

STEP 3 To extract the laser diode from its heat sink, load your soldering iron with enough solder so it will make contact with both of the diode's soldered pins. Slide a knife under the diode's ribbon, and pull up on it as you touch the solder to the diode's pins to remove the ribbon.

STEP 4 Place the laser diode in a vise and file through its heat sink. Once you've weakened the heat sink, hold it with pliers and wedge a knife under the lip of the top. Use tweezers to extract the diode from the heat sink.

STEP 5 Unscrew the AixiZ module (a laser housing unit) and place the top facedown on your workspace. Drop the

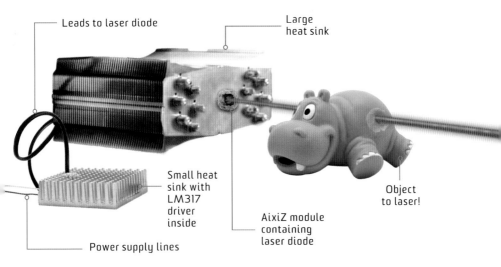

Leads to laser diode

Large heat sink

Small heat sink with LM317 driver inside

AixiZ module containing laser diode

Object to laser!

Power supply lines

laser diode in it with the laser facing down, carefully pressing it so the diode's back is flush with the module.

STEP 6 Attach two long leads to the diode's negative and positive pins and reattach the back of the AixiZ module, threading the leads through the hole in its back.

STEP 7 Solder a 3k-ohm resistor across the LM317 regulator's adjustable and output voltage pins, then solder the laser diode's positive lead to the adjustable pin on the LM317 regulator.

STEP 8 Clip the wire connectors off the ends of the wires attached to the power supply, then clip and strip the black and green wires and solder them together.

STEP 9 Solder the diode's negative lead to the power supply's red (negative) wire, and the power supply's yellow (positive) wire to the regulator's input pin.

STEP 10 Drill a ½-inch (1.25-cm) hole through the large heat sink that you harvested from your computer.

STEP 11 Slide the AixiZ module into the large heat sink, making sure it doesn't protrude. Secure with thermal glue.

STEP 12 Use thermal glue to attach the small heat sink securely to the LM317 driver you wired.

STEP 13 Put on safety goggles that are specifically designed to protect against lasers. Burn and cut things.

5 MINUTE PROJECT

059 HACK INFRARED GOGGLES

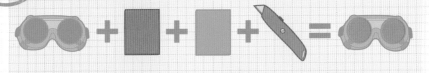

STEP 1 Unscrew the eyepieces from a pair of welding goggles. Remove the dark green welding lenses, leaving just the clear plastic.

STEP 2 Use a green welding lens as a template to cut out eight circles of blue gel sheet and two circles of red gel sheet.

STEP 3 Add four blue gel sheet pieces to each eyepiece, screw them back onto your goggles, and enjoy the crazy spectrum.

STEP 4 Add a red gel sheet piece for a different effect. Whatever you do, just don't look at the sun.

060 A PORTABLE X-RAY MACHINE

What did you do with your free time in high school? Bet you didn't build yourself a machine to see inside other machines. Adam Munich did.

Late one night, Adam Munich found himself talking with two guys online: one who complained of rolling electricity blackouts and one who had broken his leg in Mexico and said his local hospital couldn't find an X-ray machine. The two situations fused in Munich's mind; he wondered if a cheap, reliable, battery-powered X-ray machine existed—something that could be used in remote areas and function without being plugged in.

After discovering that the answer was no, he spent two years building one out of nixie tubes, old suitcases, chain-saw oil, and electronics from across the globe. It was a very ambitious project for anyone, let alone a 15-year-old. "For something like this," he said, "there's no guide."

Now that's he's figured out how to build the device, Munich's focus is on making it cheaper to recreate and sturdy enough to help his online acquaintances and others in similar situations. But the current version will no doubt help him with his immediate concern—impressing college admissions offices.

MAN AND MACHINE
Munich has used his machine to X-ray items like a pen and a computer hard drive. Theoretically, he says, it could be used for hands and limbs.

061 ILLUMINATE SKETCHES WITH HOMEMADE CONDUCTIVE INK

Write it, wire it up, and see it in lights.

MATERIALS

2.5 ml ammonium hydroxide

Two pipettes

Test tube

1 gram silver acetate

Centrifuge

0.2 ml formic acid

Syringe

Syringe filter

Glass vial

Glass to paint on

Thin paintbrush

9-volt battery

LEDs

STEP 1 Use a pipette to measure the ammonium hydroxide into a test tube, then add the silver acetate. Place the tube in the centrifuge; let it mix for 15 seconds.

STEP 2 Using a second pipette, transfer 0.2 ml formic acid into the test tube solution one drop at a time, mixing it in the centrifuge between each drop.

STEP 3 Set the test tube aside and ignore it for 12 hours.

STEP 4 Pull the plunger out of the back of a syringe and add a filter to the syringe, then decant the solution in the test tube into the syringe.

STEP 5 Open the glass vial and place it under the syringe, then place the plunger back into the syringe and force the liquid through the filter and into the vial.

STEP 6 Using the liquid in the vial, draw or write something on glass with a thin paintbrush, leaving gaps for the leads of the power source and the LEDs.

STEP 7 Heat the glass in an oven set to 200°F (93°C). Wait 15 minutes, then remove it. It should have a conductive silver coating.

STEP 8 Once you've placed the battery and LEDs on the glass, your circuit art will light right up.

LEDs

9-volt battery

Conductive ink pattern

062 A 3D PRINTER THAT RUNS ON SUN AND SAND

This bizarre-looking contraption turns the desert's resources—a whole lot of sun and sand—into glass.

When design student Markus Kayser wanted to test his sun-powered, sand-fed 3D printer, he knew the gray skies outside his London apartment wouldn't do. So he shipped the 200-pound (90-kg) device to Cairo, Egypt, hoping to find plenty of sun and sand that could, in conjunction with a large lens, produce glassware.

How does this machine work? Two aluminum arms, holding the lens at one end and solar panels at the other, can pivot from straight overhead down to a 45-degree angle to chase the sun. Sensors detect the shadows and feed the data on their position to Kayser's computer, which directs the motorized frame to adjust to properly align the lens. Two photovoltaic panels, one on either side of the machine, keep the printer powered. Since the panels are attached to the same arms as the lens, they also benefit from the sun tracking, thus ensuring that they always get direct light.

Kayser first designs the object he wants to print in a computer-assisted design (CAD) program. His computer sends instructions to the printer, which works from the bottom up. After a layer has cooled into glass, he adds more sand to the sandbox in the center of the machine and flattens it out, and the printer begins heating the next layer. Kayser's first major piece, a bowl, took about four and a half hours to print.

PRINTING A BETTER BOWL
Kayser has printed a glass bowl
and several sculptures. He admits
they're not perfect; he says he
could have used more complicated
optics. But, he adds, perfection
wasn't the point: "This is about
showing the potential."

063 BEAM A BATMAN-STYLE SPOTLIGHT

Send up a tiny desktop signal whenever you're in need of big-time assistance.

COST	$
TIME	☺ ☺
EASY ● ● ● ○ ○ HARD	

MATERIALS

Large translucent flip top bottle cap from a detergent or shampoo bottle
Craft knife
Aluminum foil
Scissors
Double-sided tape
USB laptop light
Wire cutters
Soldering iron and solder
Hot-glue gun
Clear plastic sheet
Cap from a sports bottle
Loose change
Foam board or cardboard
Black paint

STEP 1 Use a craft knife to cut off the top of the large flip-top bottle cap and discard it. Wrap the bottle cap in strips of aluminum foil, leaving small gaps between strips and adhering them with double-sided tape.

STEP 2 Cut the USB plug off the USB laptop light using the wire cutters; pull out the LED and its attached wires.

STEP 3 Flip the bottle cap over so that the exit hole is on the bottom and the cap's neck becomes the top. Crumple up aluminum foil, form it into a rough bowl shape, and place it inside the cap.

STEP 4 Punch a hole in the aluminum foil with a craft knife. Thread the LED wires through it and out the exit hole of the large cap, nestling the LED inside the bowl shape.

STEP 5 Use a craft knife to make a slit in the covering of the USB lamp's detached USB plug, and peel the covering off. Then take apart the metal case and solder the wires that once went to the LED encased inside the lamp back on to the LED's relevant pins.

STEP 6 Use hot glue to secure the wires in place. Reattach the metal case and then the rubber outer housing.

STEP 7 Cut a piece of plastic sheet to fit over the cap, with three tabs around the edge to fit down into the bottle cap. Fold down the tabs and apply glue to them, then press the plastic "lens" in place.

STEP 8 Cut a tiny logo out of aluminum foil and glue it onto the lens.

STEP 9 Use the cap from a sports bottle to make the base. Glue a couple of coins into the cap's bottom to weight it.

STEP 10 Cut a support piece out of foam board or cardboard for the spotlight to rest in. Be sure to make a slender bottom point that will fit into the small hole in the sports-bottle-cap base. Paint the base and support black.

STEP 11 Glue the foam support to the small sports-bottle-cap base, then glue the spotlight to the support.

STEP 12 Send your signal.

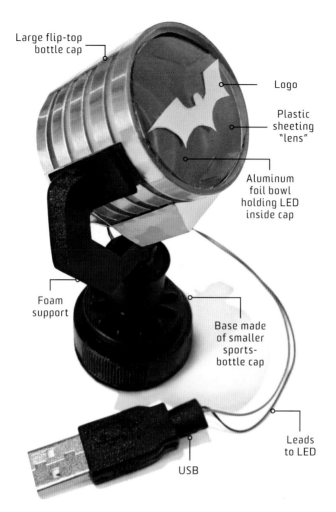

Large flip-top bottle cap

Logo

Plastic sheeting "lens"

Aluminum foil bowl holding LED inside cap

Foam support

Base made of smaller sports-bottle cap

USB

Leads to LED

5 MINUTE PROJECT

064 SHINE A MINI FLASHLIGHT

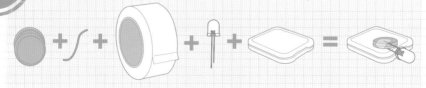

STEP 1 Connect two 3-volt batteries side by side with ¾-inch (1.9-cm) electrical wire and tape.

STEP 2 Tape the LED's leads loosely to the batteries, connecting positive to positive and negative to negative terminals.

STEP 3 Poke a hole in an SD-card case and insert the batteries with the LED terminals.

STEP 4 Squeeze the case to make the LED's leads press against the battery and light up.

065 REPURPOSE FOIL FOR TECHIE USE

Dig this stuff out of a kitchen cupboard and make the most of its conductive properties.

DO-IT-YOURSELF CAPACITORS
Cut a 2-foot (60-cm) length of aluminum foil, and three lengths of plastic wrap to match. Cut the foil in half lengthwise and tape a piece of electrical wire to each sheet, then put the pieces of foil on top of each other with plastic wrap in between. Put more plastic wrap on top of and below the foil, then roll it all up and hook the wires to a battery charger. Charge stuff up.

SECURE LOOSE BATTERIES
Sometimes, batteries don't quite connect with the springs inside your devices. So fold up a piece of aluminum foil and slide it between the battery terminals and the springs.

CELL-PHONE–SIGNAL BLOCKER
To stay under cover and prevent people from picking up your GPS coordinates, tightly wrap your phone in multiple layers of aluminum foil. To test it, try calling your phone. If it goes straight to voicemail, the signal to and from your phone is successfully blocked, and your coordinates are protected.

DIY LIGHT REFLECTOR
Wrap a large piece of cardboard in foil to create a quick and cheap light reflector for photography.

066 IMPROVISE A TRIPOD

STEP 1 Poke a hole into the center of the bottom of a plastic cup.

STEP 2 Insert a bolt that fits your camera's threaded tripod hole. Glue the bolt in place.

STEP 3 Screw on your camera and start snapping.

067 MOUNT A CAMERA TO YOUR BIKE

Document your epic rides with a bike-bell mod.

MATERIALS

Bicycle bell
Camera

Bicycle
Screwdriver

STEP 1 Find a bike bell with a central screw that fits the tripod mount on the bottom of your camera. Most tripod mounts measure ¼ inch (6.35 mm).

STEP 2 Attach the bell to the handlebars.

STEP 3 Use a screwdriver to remove the bell's dome.

STEP 4 Screw the camera's tripod mount to the bell's central screw. Orient the camera whichever way you like, and start shooting your photographic travelogue.

Camera

Central screw

Bicycle bell

068 BUILD A TIME-LAPSE CAMERA STAND

Upgrade your kitchen timer for slick panoramic photos on the cheap.

MATERIALS

Kitchen timer
Drill
1/4-inch (6.35-mm) 20 set screw
3/8-inch (9.5-mm) bolt
3/8-inch (9.5-mm) 20 bushing
Craft knife
Rubber mat
Glue
Tripod
Tripod mount
Camera
Computer with photo-editing software

STEP 1 Drill a 15/64-inch (6-mm) hole into the center of a kitchen timer's dial. Insert a 1/4-inch (6.35-mm) 20 set screw.

STEP 2 In the bottom of the timer, drill a hole that measures 11/32 inches (8.75 mm) in diameter. Screw a 3/8-inch (9.5-mm) bolt into the hole to create threads for a 3/8-inch (9.5-mm) bushing (a threaded insert that will allow you to mount the timer to the tripod).

STEP 3 Measure and cut a piece of rubber mat, leaving a hole for the bushing, and glue it to the bottom of the timer.

STEP 4 Mount the timer on a tripod, attach the tripod mount and camera, and set the timer. Adjust the camera's settings to take pictures at regular intervals, and then transfer the shots to a computer and create a panoramic time-lapse montage with photo software.

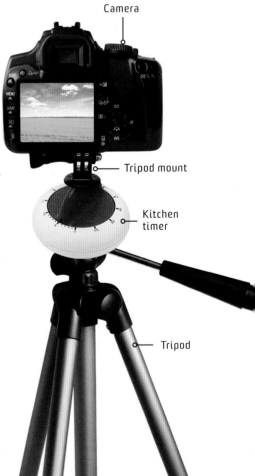

Camera

Tripod mount

Kitchen timer

Tripod

069 Snap a Self-Portrait with a DIY Remote Shutter Release

Trigger your camera's shutter from afar with basic household parts.

MATERIALS

Rotary tool
Pill bottle with snap-on lid
Craft knife
Rubber tubing that fits
  snugly over the nozzle of
  your squeeze bottle
Shutter release cable
Plastic squeeze bottle
Short piece of plastic tubing
  that fits snugly into the
  rubber tubing
Balloon
Thread
Talcum powder
Cork that fits into the pill bottle

STEP 1 Use a rotary tool to drill a hole into the pill bottle's lid big enough to fit the rubber tubing.

STEP 2 Drill a hole in the bottom of the pill bottle where you'll thread your shutter release cable. Then insert the trigger.

STEP 3 Use a craft knife to cut the rubber tubing long enough to cover the longest distance that you anticipate being from the camera. Insert the nozzle of your squeeze bottle into one end of the rubber tubing.

STEP 4 Thread the tubing's other end through the hole in the pill bottle's lid and then onto the short plastic tubing.

STEP 5 Pull the balloon's opening around the piece of plastic tubing and tie it on securely with thread.

STEP 6 Dust the piece of cork with talcum powder and insert it into the pill bottle. Lower the balloon into the bottle so that it rests against the cork.

STEP 7 Test that everything is airtight, then hook the cable to your camera and get in front of the camera, holding the squeeze bottle. When you're ready, squeeze the bottle, and the balloon will inflate, pushing the cork against the cable's trigger. Say "cheese," anyone?

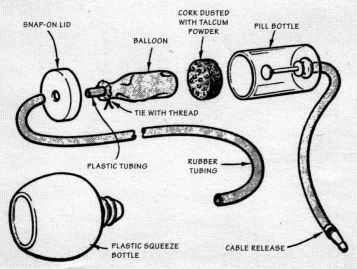

SNAP-ON LID

BALLOON

CORK DUSTED
WITH TALCUM
POWDER

PILL BOTTLE

TIE WITH THREAD

PLASTIC TUBING

RUBBER
TUBING

PLASTIC SQUEEZE
BOTTLE

CABLE RELEASE

070 RIG A PLASTIC-BOTTLE DIFFUSER

STEP 1 Cut out a small section of an empty frosted plastic water jug.

STEP 2 Cut a hole in the section large enough for the camera's lens to fit through.

STEP 3 Place the diffuser on the camera so it covers the flash. Snap away.

071 MAKE YOUR CAMERA WATERPROOF

STEP 1 Cut a piece of toilet-paper tube to match the depth of your camera lens and cover your lens with it.

STEP 2 Stretch an unlubricated condom open. Add a packet of desiccant gel—it will absorb moisture—and slide your camera inside with the tube in place.

STEP 3 Tie the condom slack into a knot and superglue the knot to make it watertight.

STEP 4 Stretch a second condom open and insert the wrapped camera, knot side in. Tie and glue the knot again. Dive in and document.

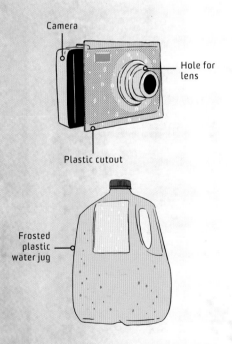

Camera

Hole for lens

Plastic cutout

Frosted plastic water jug

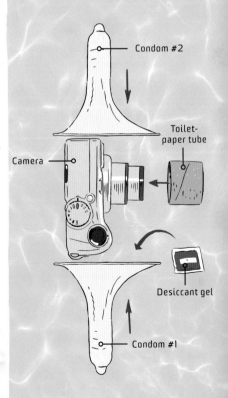

Condom #2

Toilet-paper tube

Camera

Desiccant gel

Condom #1

072 CREATE A PEEPHOLE FISHEYE LENS

STEP 1 Turn on your camera and extend the zoom as far as it will go.

STEP 2 Grab yourself a standard peephole from a home-supply store and, making sure the peephole is facing the right way out, pop it over the camera lens.

STEP 3 Hold it in place or attach it to the lens with heavy-duty tape. (Look for tape that doesn't leave a residue, which could make your lens stick.)

073 ADAPT A MANUAL LENS TO YOUR DSLR

STEP 1 Procure a twist-on adapter ring and a compatible manual-focus lens from a vintage film camera. (These options are both way less expensive than the lenses you can buy for a DSLR, and you can even find old macro, fisheye, and ultrazoom lenses relatively inexpensively—just make sure that they're compatible with your DSLR before purchasing.)

STEP 2 Insert the lens into the adapter. There is usually a dot or some kind of marking on both the lens and the adapter that makes it easy to see how the two fit together.

STEP 3 Twist the lens while holding the adapter in place, as if you were mounting the lens onto a camera body. The lens should make a click or locking sound when it's secured correctly.

STEP 4 Mount the lens and adapter combination onto your camera like any other lens and get shooting.

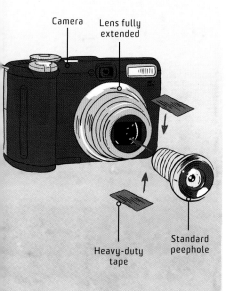

Camera

Lens fully extended

Heavy-duty tape

Standard peephole

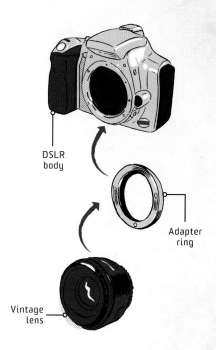

DSLR body

Adapter ring

Vintage lens

074 SET UP A HIGH-SPEED AUDIO-TRIGGERED FLASH

Light up superquick action with a superquick audio-triggered flash.

COST	$$
TIME	☺ ☺ ☺
EASY	● ● ● ● ○ HARD

MATERIALS

Disposable flash camera

Two AA and two AAA alkaline batteries

Voltage meter

Sensitive-gate, 400 volt, 0.8-amp silicon-controlled rectifier (SCR)

Wire strippers

Soldering iron and solder

3.5-mm stereo cable

Cassette recorder

Electret microphone

Camera capable of "B" (bulb) or prolonged exposures (at least two seconds)

Tripod

STEP 1 Remove the exterior plastic, film advance system, and shutter assembly from the disposable flash camera. Make sure it has fresh batteries.

STEP 2 Locate the camera's flash terminals, located near where the shutter was. Use a voltage meter to determine which terminals is positive and which is negative.

STEP 3 Solder the cathode pin of the silicon-controlled rectifier to the negative flash terminal and the anode pin to the positive flash terminal.

STEP 4 Snip off one jack of the 3.5-mm stereo cable. Peel back its plastic to expose the red, white, and ground wires.

STEP 5 Solder the stereo cable's red and white wires to the silicon-controlled rectifier's gate pin and its ground wire to the negative flash terminal (with the cathode pin).

STEP 6 Plug an inexpensive electret microphone into the tape recorder's mic input. Plug the jack of the 3.5-mm stereo cable into the audio output.

STEP 7 Remove the tape recorder's door and look for the recorder's write-protection button, a small movable "finger" opposite the record head. Hold down this button and press the recorder's red record button to start charging the disposable camera's flash. When the amber ready light glows steadily, you can start using it.

STEP 8 Find a dark area, mount your camera onto a tripod, and set the shutter of the camera you'll be shooting with for a bulb exposure. (Alternatively, cameras that can deliver timed 1- to 4-second exposures can work.)

STEP 9 Kill the lights, hold the flash trigger near your subject, open the camera's shutter, and record a high-speed event that is accompanied by a noise—such as the pop of a water balloon or the smack of a slap. Take your pick and take some pics.

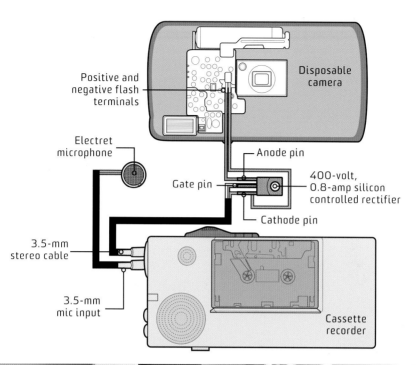

Positive and negative flash terminals

Disposable camera

Electret microphone

Anode pin

Gate pin

400-volt, 0.8-amp silicon controlled rectifier

Cathode pin

3.5-mm stereo cable

3.5-mm mic input

Cassette recorder

Photo of a noisy, fast event

075 A CAMERA THAT SHOOTS HUGE PHOTOGRAPHS

This beast of a camera allows you to capture huge images on X-ray film. Problem is, you'll have to build it first.

Darren Samuelson had just taken his last photo of Manhattan when the police arrived. He and his father had been working from an empty dock across the Hudson River, and the authorities wanted to know what they were doing with a folding contraption that was more than 6 feet (1.8 m) long and 70 pounds (32 kg) pointed at the city. Samuelson pleaded that it was a camera, and that he was just a tourist. They believed him and he got his shot—a photo so detailed that the print could be blown up to half the length of a volleyball court and still remain sharp.

Samuelson specially built this camera for X-ray stock that measures 14 by 36 inches (35 cm by 90 cm) and is cheaper than large-format photo paper. He began by constructing the massive accordionlike bellows required to adjust the camera's focal length manually, spending two weeks on the floor folding, cutting, gluing, and inserting the ribs that would give it form. The camera and bellows unfold and slide out on rails, with a lens at one end and the film holder at the other. To focus, he slides either end in or out. The result is not point-and-shoot, Samuelson admits, and the build wasn't easy (the parts list runs to 186 rows on a spreadsheet). "But when I hold up a print and see the amazing detail," he says, "I think, 'Yeah, this was worth it.'"

MEGA PRINTS
Each print measures 3 feet (90 cm), and while shooting Samuelson drapes an immense black cloak over himself and the camera to block out light.

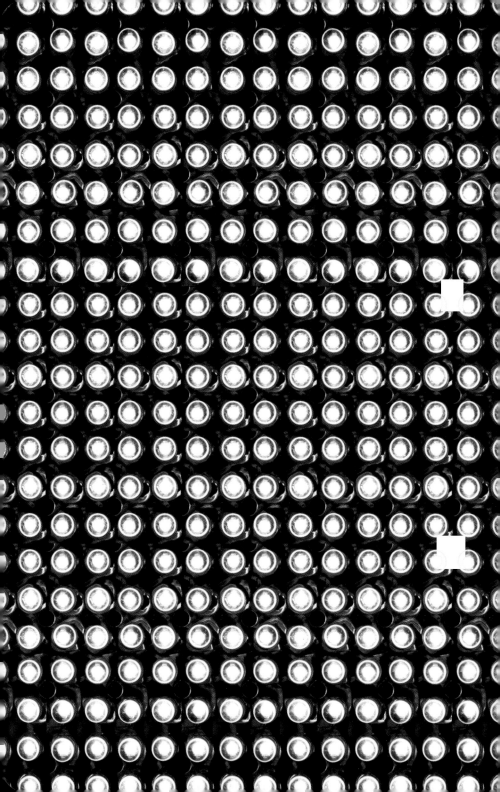

RESOURCES

GLOSSARY

ALLIGATOR CLIP Spring-loaded clip that can be used to connect a component to a wire in a temporary circuit.

AMPLIFIER Component that augments the power of a signal. In circuits, an amplifier is used to increase voltage or current.

ANTENNA Wire, thin metal pole, or other device that can transmit or receive electromagnetic waves, such as TV or radio waves.

ARDUINO Common, open-source microcontroller. There are various types of Arduino microcontrollers that use the same programming language.

BREADBOARD Base used to test temporary circuits before soldering them.

BREAKOUT BOARD Electrical component that allows easier access to tightly spaced pins on a microchip or to densely bundled wires.

BUSHING Connector used to join pipes of different diameters; one end has a smaller opening, the other end a larger one. A bushing can also be called a reducing coupling.

CAPACITOR Electrical component that stores energy within a circuit. Unlike a battery, a capacitor does not produce energy; it contains or filters the energy already flowing through the circuit.

CIRCUIT Closed loop through which electrical current flows. A circuit is often used to power an electrical device.

CIRCUIT BOARD Thin, insulated board on which electrical components are mounted and connected together.

CLAMP Device used to hold an object tightly in place. Clamps can vary widely and can be intended for temporary or permanent use.

COAXIAL CABLE Cable with a central conductive wire, surrounded by an insulating layer, which in turn is surrounded by a conductive tube. A coaxial cable is used to transmit radio or cable signals.

CONDUCTIVITY Capacity to transmit an electrical current; it can also refer to the measure of a substance's ability to transmit electrical current.

CONTACT Point where a circuit component is connected to a wire or circuit board.

CRAFT KNIFE Small, fixed-blade knife used to make precise cuts.

DESOLDERING Removing solder to detach components from a circuit or circuit board.

DIODE Electronic component with one terminal that has high resistance, and another terminal with low resistance. A diode is used to allow current to flow in one direction but not another.

DRILL Tool used to cut holes in a variety of materials. A drill is usually powered by electricity, and comes with variously sized bits.

ELECTRICAL TAPE Type of tape covered in an insulating material, often used to cover and connect electrical wires.

ELECTRICAL WIRE Insulated strand of conductive material used to carry electricity.

ELECTRODE Conductor used to transmit current to a non-metallic material. Electrodes are used in arc welding.

EPOXY Adhesive made from a type of resin that becomes rigid when heated or cured.

FIBER OPTIC CABLE Cable made up of thin fibers that transmit light from one end to the other.

FLASH DRIVE Small data-storage device that can be connected to a computer, often via a USB port.

FRESNEL LENS Thin lens made of a number of smaller lens segments. A Fresnel lens magnifies a light source, and is often used in projectors and spotlights.

GROUND WIRE Wire in a circuit that provides a return path for current, often into the earth.

HACKSAW Fine-toothed saw held in a frame. A hacksaw can be used to cut metal or other hard materials.

HEAT-SHRINK TUBING Tubing that contracts when heated; often used to insulate wires or to create a protective seal.

HEAT SINK Device that channels heat away from an electronic system, keeping it cool enough to operate properly.

INSULATION Material, such as the coating around an electrical wire, that prevents current or heat from flowing.

INTEGRATED CIRCUIT Also called a microchip, this device allows complex circuitry to be condensed into an extremely small space.

JIGSAW Tool (usually a power tool) with a long, thin saw blade. A jigsaw is useful in cutting curves and irregular shapes.

LASER Device that emits a tightly focused beam of light.

Lasers vary in intensity, and can be strong enough to cut through tough materials.

LCD MONITOR Display that uses liquid crystals (sandwiched between two electrodes and filtered through polarizing film) to create an image.

LEAD A wire extending from an electronic component that is used to connect that component to another electronic part.

LED Diode that gives off light. They are usually more energy-efficient than incandescent light sources, and usually smaller.

LITHIUM-ION BATTERY Type of rechargeable battery that is often used in consumer electronics and carries a large amount of energy for its size.

MICROCONTROLLER Tiny dedicated computer, contained on a single chip, that can be embedded in a larger device.

MULTIMETER Device that measures electrical current, resistance, and voltage, and is helpful for monitoring circuits.

OHM Unit of measurement of electrical resistance.

OPTICAL MOUSE Computer mouse that uses an LED to sense motion.

PHOTOCELL A device that detects the presence (or absence) of light and adjusts current flow accordingly.

PLEXIGLAS Hard, transparent plastic; looks like glass but is more lightweight and durable.

PORT Point of interface between two devices.

POTENTIOMETER Three-terminal electrical component that acts as a variable resistor.

These adjust the flow of current through a circuit, and are often used in dimmer switches.

PROGRAMMING LANGUAGE Language used to convey instructions to a computer or other machine. Many distinct programming languages are used for different purposes.

PROJECT BOX Box designed to contain the components of a circuit; useful for mounting and protecting a circuit's elements.

PVC PIPE Type of durable, lightweight plastic pipe.

PVC PIPE CEMENT Adhesive designed to connect pieces of PVC material together.

REBAR Ridged bar of steel, often used in construction to reinforce concrete or masonry.

REED SWITCH Switch composed of metal reeds enclosed in a tiny glass container. The reeds react to the presence of a magnetic field.

RESISTOR Two-terminal electrical component that resists the flow of an electric current. A resistor is used in a circuit to control the direction and strength of current.

ROTARY TOOL Power tool with a wide variety of interchangeable bits that can cut, polish, carve, or grind.

SAFETY GOGGLES Glasses that shield the eye area from heat, chemicals, and debris.

SCHEMATIC Two-dimensional map of an electrical circuit that uses a set of symbols to stand for the components of a circuit, and shows their connections.

SKETCH When working with Arduino microcontrollers, a

sketch is a program that can be loaded into an Arduino.

SOLDERING Connecting two metal objects together by melting solder (a type of metal) to create a strong joint.

SUGRU Type of silicone-based putty that cures to a solid but flexible state after 24 hours.

SWITCH Component that can stop the flow of current in a circuit, or allow it to continue.

TABLE SAW Machine that cuts wood or other materials with a spinning serrated metal disc.

TERMINAL End point of a conductor in a circuit; also a point at which connections can be made to a larger network.

TOUCHSCREEN An electronic screen that users interact with by touching it with their fingers.

TRANSFORMER Device that transfers current from one circuit to another or alters the voltage of a current.

TRANSISTOR Semiconducting electrical component with at least three leads; can control or amplify the flow of electricity.

USB Universal Serial Bus; a common type of connector for computers and electronics.

VISE Type of clamp, often affixed to a table, that uses a screw to hold an object in place.

VOLT Unit of measurement for electrical potential.

WIRE STRIPPER Device composed of a set of scissor-like blades with a central notch; used to strip the insulation from the outside of electrical wires.

ZIP TIES Self-closing fastener that creates a secure loop.

INDEX

THANKS TO OUR MAKERS

Lots of inventive people contributed their ideas and how-to tutorials to the pages of this book. Look them up to find out more details about their projects, as well as any new cool stuff they're up to.

TECH UPGRADES

007: William Finucane (adapted with permission from his original guide in the Mad Science World on wonderhowto.com) **012:** Wallace Kineyko **013:** Dave Prochnow **017:** Dylan Hart (householdhacker.com) **021:** Jennifer Lee (jen7714.wordpress.com) **022:** Daniel Julian **026:** Adapted from *More Show Me How* **027:** Windell H. Oskay (evilmadscientist.com) **029:** Instructables username unclesam **031:** Jani "Japala" Pönkkö (editor of metku.net) **032:** Ian Cannon **033:** Jared Bouck (inventgeek.com) **034:** Dave Fortin (failsworld.com) **035:** Ian Cannon **038:** Phil Herlihy (braindeadlock.net) **039:** Jeffrey Davies **041:** Bard Lund Johansen **045:** Toma Dimov (outfab.com) **047:** Ingo Schommer (chillu.com) **049:** Pamela Stephens (pbjstories.com) **055:** Bionerd23 (youtube.com/user/bionerd23) **056:** Ben Krasnow (youtube.com/user/bkraz333) **057:** Burak Incepinar (tacashi.tripod.com) **059:** Kip Kay (kipkay.com) **060:** Adam Munich **062:** Markus Kayser (markuskayer.com) **063:** Chris Barnardo **068:** Larry Towe (getawaymoments.com) **074:** Dave Prochnow **075:** Darren Samuelson (darrensamuelson.com)

IMAGE CREDITS

DISCLAIMER

The information in this book is presented for an adult audience and for entertainment value only. While every piece of advice in this book has been fact-checked and where possible, field-tested, much of this information is speculative and situation-dependent. The publisher assumes no responsibility for any errors or omissions and makes no warranty, express or implied, that the information included in this book is appropriate for every individual, situation, or purpose. Before attempting any activity outlined in these pages, make sure you are aware of your own limitations and have adequately researched all applicable risks. This book is not intended to replace professional advice from experts in electronics, woodworking, metalworking, or any other field. Always follow all manufacturers' instructions when using the equipment featured in this book. If the manufacturer of your equipment does not recommend use of the equipment in the fashion depicted in these pages, you should comply with the manufacturer's recommendations. You assume the risk and full responsibility for all of your actions, and the publishers will not be held responsible for any loss or damage of any sort—whether consequential, incidental, special, or otherwise—that may result from the information presented here. Otherwise, have fun.

weldon**owen**

President, CEO Terry Newell

VP, Sales Amy Kaneko

VP, Publisher Roger Shaw

Senior Editor Lucie Parker

Project Editors Emelie Griffin, Jess Hemerly

Creative Director Kelly Booth

Designer Michel Gadwa

Image Coordinator Conor Buckley

Production Director Chris Hemesath

Production Manager Michelle Duggan

415 Jackson Street, Suite 200
San Francisco, CA 94111
Telephone: 415 291 0100
Fax: 415 291 8841
www.weldonowen.com

Popular Science and Weldon Owen are divisions of **BONNIER**

Library of Congress Control Number is on file with the publisher.

ISBN 10: 1-61628-531-1
ISBN 13: 978-1-61628-531-9

10 9 8 7 6 5 4 3 2 1
2013 2014 2015

Printed in China by 1010 Printing.

POPULAR SCIENCE ⊕ THE FUTURE NOW

ACKNOWLEDGMENTS

Weldon Owen would like to thank Jacqueline Aaron, Katie Cagenee, Andrew Jordon, Katharine Moore, Gail Nelson-Bonebrake, Jenna Rosenthal, Katie Schlossberg, and Marisa Solís for their editorial expertise and design assistance.

We'd also like to thank our technical editors, Michael Rigsby and Tim Lillis, and our in-house builder and circuitry diagram consultant, Ian Cannon.

Popular Science would like to thank Matt Cokeley, Todd Detwiler, Kristine LaManna, Stephanie O'Hara, Thom Payne, and Katie Peek for their support over the years.

We would also like to thank Gregory Mone for penning the You Built What?! entries included in this book.

And a big thanks to Mark Jannot and Mike Haney—the How 2.0 column's first editor—for getting it all started.